Nanostructured Surfaces and Thin Films Synthesis by Physical Vapor Deposition

Nanostructured Surfaces and Thin Films Synthesis by Physical Vapor Deposition

Editor

Rafael Alvarez

MDPI • Basel • Beijing • Wuhan • Barcelona • Belgrade • Manchester • Tokyo • Cluj • Tianjin

Editor
Rafael Alvarez
Instituto de Ciencia de Materiales de Sevilla (ICMS)
Departamento de Física Aplicada I. Escuela Politécnica Superior,
Universidad de Sevilla
Spain

Editorial Office
MDPI
St. Alban-Anlage 66
4052 Basel, Switzerland

This is a reprint of articles from the Special Issue published online in the open access journal *Nanomaterials* (ISSN 2079-4991) (available at: https://www.mdpi.com/journal/nanomaterials/special_issues/nano_surf_thin_phys).

For citation purposes, cite each article independently as indicated on the article page online and as indicated below:

LastName, A.A.; LastName, B.B.; LastName, C.C. Article Title. *Journal Name* **Year**, *Volume Number*, Page Range.

ISBN 978-3-0365-0394-3 (Hbk)
ISBN 978-3-0365-0395-0 (PDF)

© 2021 by the authors. Articles in this book are Open Access and distributed under the Creative Commons Attribution (CC BY) license, which allows users to download, copy and build upon published articles, as long as the author and publisher are properly credited, which ensures maximum dissemination and a wider impact of our publications.

The book as a whole is distributed by MDPI under the terms and conditions of the Creative Commons license CC BY-NC-ND.

Contents

About the Editor . . . . **vii**

Alberto Palmero, German Alcala and Rafael Alvarez
Editorial for Special Issue: Nanostructured Surfaces and Thin Films Synthesis by Physical Vapor Deposition
Reprinted from: *Nanomaterials* **2021**, *11*, 148, doi:10.3390/nano11010148 1

Adriano Panepinto and Rony Snyders
Recent Advances in the Development of Nano-Sculpted Films by Magnetron Sputtering for Energy-Related Applications
Reprinted from: *Nanomaterials* **2020**, *10*, 2039, doi:10.3390/nano10102039 7

Ho-young Jeong, Seung-hee Nam, Kwon-shik Park, Soo-young Yoon, Chanju Park and Jin Jang
Significant Performance and Stability Improvements of Low-Temperature IGZO TFTs by the Formation of In-F Nanoparticles on an SiO_2 Buffer Layer
Reprinted from: *Nanomaterials* **2020**, *10*, 1165, doi:10.3390/nano10061165 33

Chzu-Chiang Tseng, Gwomei Wu, Liann-Be Chang, Ming-Jer Jeng, Wu-Shiung Feng, Dave W. Chen, Lung-Chien Chen and Kuan-Lin Lee
Effects of Annealing on Characteristics of $Cu_2ZnSnSe_4/CH_3NH_3PbI_3/ZnS/IZO$ Nanostructures for Enhanced Photovoltaic Solar Cells
Reprinted from: *Nanomaterials* **2020**, *10*, 521, doi:10.3390/nano10030521 41

Asma Chargui, Raya El Beainou, Alexis Mosset, Sébastien Euphrasie, Valérie Potin, Pascal Vairac and Nicolas Martin
Influence of Thickness and Sputtering Pressure on Electrical Resistivity and Elastic Wave Propagation in Oriented Columnar Tungsten Thin Films
Reprinted from: *Nanomaterials* **2020**, *10*, 81, doi:10.3390/nano10010081 57

Eugen Stamate
Spatially Resolved Optoelectronic Properties of Al-Doped Zinc Oxide Thin Films Deposited by Radio-Frequency Magnetron Plasma Sputtering Without Substrate Heating
Reprinted from: *Nanomaterials* **2020**, *10*, 14, doi:10.3390/nano10010014 75

Rafael Alvarez, Sandra Muñoz-Piña, María U. González, Isabel Izquierdo-Barba, Iván Fernández-Martínez, Víctor Rico, Daniel Arcos, Aurelio García-Valenzuela, Alberto Palmero, María Vallet-Regi, Agustín R. González-Elipe and José M. García-Martín
Antibacterial Nanostructured Ti Coatings by Magnetron Sputtering: From Laboratory Scales to Industrial Reactors
Reprinted from: *Nanomaterials* **2019**, *9*, 1217, doi:10.3390/nano9091217 87

Yuan-Chang Liang and Yen-Chen Liu
Design of Nanoscaled Surface Morphology of TiO_2–Ag_2O Composite Nanorods through Sputtering Decoration Process and Their Low-Concentration NO_2 Gas-Sensing Behaviors
Reprinted from: *Nanomaterials* **2019**, *9*, 1150, doi:10.3390/nano9081150 101

Lingwei Ma, Jinke Wang, Hanchen Huang, Zhengjun Zhang, Xiaogang Li and Yi Fan
Simultaneous Thermal Stability and Ultrahigh Sensitivity of Heterojunction SERS Substrates
Reprinted from: *Nanomaterials* **2019**, *9*, 830, doi:10.3390/nano9060830 115

Jonathan Colin, Andreas Jamnig, Clarisse Furgeaud, Anny Michel, Nikolaos Pliatsikas, Kostas Sarakinos and Gregory Abadias
In Situ and Real-Time Nanoscale Monitoring of Ultra-Thin Metal Film Growth Using Optical and Electrical Diagnostic Tools
Reprinted from: *Nanomaterials* **2020**, *10*, 2225, doi:10.3390/nano10112225 . . . . **125**

Xiaoyu Zhao, Jiahong Wen, Aonan Zhu, Mingyu Cheng, Qi Zhu, Xiaolong Zhang, Yaxin Wang and Yongjun Zhang
Manipulation and Applications of Hotspots in Nanostructured Surfaces and Thin Films
Reprinted from: *Nanomaterials* **2020**, *10*, 1667, doi:10.3390/nano10091667 . . . . **155**

About the Editor

Rafael Alvarez (Ph.D.) received his Ph.D. degree in July 2005 from the Physics Department of the University of Cordoba (Spain). He has been a member of the Materials Science Institute of Seville since 2008, and he has recently joined the Applied Physics Department of the University of Seville as an Associate Professor. His research focuses on the study and control of nanostructuring processes during thin film growth by physical vapor deposition techniques, both at the surface and at the gas phase. He is an expert on oblique angle deposition by magnetron sputtering and Monte Carlo simulations of thin film growth. His research activities include fundamental studies on the effect of the thermalization of the species arriving at the surface, the dependence of the thin film nanostructure on the material of the film or on the energy of the arriving particles, or the effect of substrate patterns on the deposited thin film features. His research has been applied to the development of antibacterial surfaces, cholesterol sensors, optical sensors, ultrasmooth surfaces, or plasmonic black metal coatings. He has registered three patents and one software application. He won the outstanding young researcher award from the Spanish Scientists Association, and the Manuel Losada Villasante prize for innovation in research. He has published 57 scientific papers, with over 750 citations, and has a h-index of 17.

Editorial

Editorial for Special Issue: Nanostructured Surfaces and Thin Films Synthesis by Physical Vapor Deposition

Alberto Palmero [1], German Alcala [2] and Rafael Alvarez [1,3,*]

[1] Materials Science Institute of Seville (CSIC-US), Américo Vespucio 49, 41092 Seville, Spain; alberto.palmero@csic.es
[2] Departamento Ingeniería Química y Materiales, Universidad Complutense de Madrid, Avenida Complutense s/n, Facultad de Ciencias Químicas, 28040 Madrid, Spain; galcalap@ucm.es
[3] Departamento de Física Aplicada I. Escuela Politécnica Superior, Universidad de Sevilla, c/Virgen de África 7, 41011 Seville, Spain
* Correspondence: rafael.alvarez@icmse.csic.es

Received: 21 December 2020; Accepted: 5 January 2021; Published: 9 January 2021

The scientific interest in the growth of nanostructured surfaces and thin films by means of physical vapor deposition (PVD) techniques has undoubtedly increased in the last decade [1]. Even though some of them can be considered mature, as they were first implemented and analyzed a few decades ago [2], the progressive understanding of fundamental atomistic phenomena [3,4], as well as their popularity in the technological industry, have prompted the scientific community to propose novel growth methodologies/geometries which, based on the classic approaches, have bestowed an even larger versatility in terms of nanostructural possibilities. In this way, film nanostructures that were unthinkable years ago are now possible by PVD: for instance, a classical technique such as the magnetron sputtering (MS) deposition method, that originally aimed to produce highly compact coatings, is currently being explored for the production of highly porous thin films thanks to the oblique angle deposition geometry [5,6]. Moreover, an MS-based methodology named High Power Impulse Magnetron Sputtering (HiPIMS) is employing high power pulsed electromagnetic signals to produce highly compact films, expanding the classic methodology [7]. Hence, it can be said that the research and application of PVD techniques are living a scientific golden age due to the large amount of exciting new possibilities, fundamental discoveries, and the numerous potential applications based on their wide range of morphological characteristics and properties, e.g. in photovoltaic cells [8], tribological coatings [9,10], optofluidic sensors [11], energy storage [12], etc.

Unfortunately, the processes responsible for the formation of a certain nanostructure by PVD are, in most of the cases, not yet known with the sufficient depth to gain enough level of control to optimize its functionality when incorporated into devices. It is then necessary to study novel nanostructures and their properties as well as the nanostructuration mechanisms, by addressing fundamental issues to understand them. In this special number, ten research works analyze various relevant aspects concerning new approaches and methodologies based on classic PVD techniques to produce films with singular nanostructures and morphologies, representing the current state of the art in this research field. Among them, two important review papers are included allowing any potential reader to understand the most relevant advances in the following two different topics. In ref. [13] the growth of metallic ultra-thin films by continuous dynamic monitoring is reviewed in depth, by combining in situ and real-time optical and electrical probes to analyze the first stages of growth in the MS deposition of a large number of metals with different crystalline structures. In ref. [14], a review on the growth of nanostructured surfaces with plasmonic properties, obtained by the combination of nanosphere lithography and PVD, is presented. There, the authors analyze the recent advances linking the fabrication routes, the film nanostructure and its plasmonic properties, with special emphasis on its application for the early detection of hepatocellular carcinoma using surface-enhanced Raman scattering and controlling the growth of Ag nanoparticles.

In addition to these relevant review manuscripts, two original research works are presented [15,16]. In ref. [15] the authors study nanostructured thin films grown by MS deposition at oblique angles, in this case using tungsten. These films are characterized by a porous structure consisting of tilted nanocolumns. By varying the deposition conditions, the tilt angle, thickness and separation of these columns can be affected, which in turn affects the optical, electrical or mechanical properties of the film. Here, the authors study the effect of both the film thickness and the gas pressure within the deposition chamber on properties such as the electrical resistivity of the film, or the anisotropy of the propagation of 2-dimensional elastic waves along the film. They have found that anisotropy rises at low pressures, where the columnar structures are better defined and more tilted, and that it increases with the thickness of the film. Electrical resistivity was found to decrease with the film thickness and to increase with the gas pressure, due to changes in the crystallinity of the film. A thorough characterization of AZO (Al-doped zinc oxide) thin films deposited by radio frequency (RF) MS is covered in ref. [16]. Sheet resistance, thin film thickness, resistivity, hall mobility, carrier concentration, optical transmittance, and band gap energy were determined for these transparent and conducting thin films as a function of the substrate position within the substrate holder with a 3 mm spatial resolution, for varying conditions of pressure, applied power, and substrate-target distance. The results show a great resistivity dependence with the substrate position, varying about 2 orders of magnitude within a distance of 10 mm. Contrarily, the spatial profile of the transmittance is quite homogenous along the surface of the samples. A reduction of the energetic oxygen ions coming from the erosion track of the target is proposed as a way to produce more homogenous films.

Other works included here focus on fundamental analysis, while also targeting a specific application, such as sensors [17,18], solar cells [19], or biomedicine [20]. In ref. [17] the authors synthesize composite TiO_2-Ag_2O nanorods to be used as chemoresistive sensors for the detection of trace amounts of NO_2 gas. TiO_2 nanorods were first prepared by hydrothermal methods, and then RF-MS of an Ag target in presence of oxygen was used to deposit Ag_2O on the TiO_2 nanorods with different degrees of coverage. The gas-sensing performance of the composite nanorods, when the coverage was made in the form of discrete Ag_2O particles, was found to be superior to that of pristine TiO_2 nanorods and even other TiO_2-Ag_2O sensors previously reported in the literature. Contribution [18] presents a novel technique that improves Ag nanorods substrates used for surface-enhanced Raman scattering (SERS), rising their sensibility by four times and increasing their thermal stability range by more than 100 °C. The technique consists of the deposition of an ultrathin capping Al_2O_3 layer on the top of the Ag nanorods, followed by the deposition of an additional Ag capping layer on top of the previous one. The Al_2O_3 layer improves thermal stability, while the last Ag layer boosts the SERS sensitivity. The Ag nanorods and the capping layers were grown using the electron-beam evaporation deposition technique at oblique angles. A detailed study of the growth of nano-sculpted thin films by the MS technique in glancing angle deposition configuration is displayed in ref. [19]. The effects of key deposition parameters, such as the deposition angle, the gas pressure within the reactor, the temperature of the substrate or the way the substrate is rotated during deposition are experimentally studied, and found to be in agreement with numerical simulations of the film growth. Film properties such as the tilt angle of the grown nanocolumns, the film porosity and, in some cases, its composition and crystalline phase, are also characterized. Finally, the authors propose the integration of nano-sculpted TiO_2 coatings into the photo-anode of dye-sensitized solar cells, concluding that a hybrid system incorporating both nanocolumns and nanoparticles significantly improves their efficiency. In ref. [20] a novel experimental methodology to produce porous thin films by MS on large surfaces is tested as an alternative to typical oblique angle deposition geometries. For this, two-side implant plates with areas up to 15 cm^2 were coated with Ti nanocolumns using an industrial reactor. While this was already achieved on small surfaces using laboratory reactors, the relevance of this paper resides on the development of a new geometrical approach to achieve the oblique incidence on industrial reactors as well as the homogeneous coating of a large area plate with Ti nanocolumns. Moreover, the authors demonstrate that the functionality of the obtained porous Ti coatings is maintained, exhibiting the same antibacterial properties as those produced at the laboratory.

Finally, important applications based on these films are developed for thin film transistors [21] or photovoltaic solar cells [22]. In ref. [21] the authors describe a way to improve the performance and stability of indium–gallium–zinc–oxide (IGZO) thin-film transistors (TFTs), which are widely used in active-matrix displays. By pretreating the substrate (a SiO_2 buffer layer) with F-plasma in a reactive ion etching chamber before depositing a 30 nm IGZO layer by magnetron sputtering, they describe the formation of indium fluoride nanoparticles in the interface, which increases the density of the IGZO, thus improving mobility and bias stability of these oxide TFTs, and allowing for their fabrication at lower temperatures. In order to improve the characteristics and efficiency of a photovoltaic solar cell, in ref. [22] the authors introduce a $Cu_2ZnSnSe_4$ (CZTSe) nano-layer between the metallic (Mo) electrode contact layer and the active absorbing perovskite layer ($MAPbI_3$). This nano-layer was deposited by RF-MS and was later subjected to an annealing process, which greatly improved its hole mobility and its role as heterojunction layer.

As derived from the ten works presented in this special number, the PVD technology comprises a family of several deposition techniques, each one of them embracing a number of different geometric configurations and deposition parameters with a direct effect on the nanostructure, morphology and properties of the produced film. It also allows deposition of most metals and many ceramic materials on a wide variety of substrates, since the deposition temperature is low enough to avoid modifications during the process on most substrate material candidates. Additionally, this is a technology with an important presence in several industrial sectors, which confirms that many scalability and production issues are already overcome. All these features provide the possibility of new developments in a huge range of possible applications, some of them already broadening like among others solar cell components, electronic and photonic applications, gas, liquid and pressure sensors and actuators, biosensors, cell-surface interaction, piezoelectric nanogenerators, electrochromic applications, water splitting fuel cells and hydrogen storage, Li-ion batteries, photovoltaic applications, surface controlled wettability, nanocarpet effect, anisotropic wetting, etc.

However, in order to reach a given surface nanostructure and architecture, and its consequent set of desired properties, a deep understanding at an atomistic level of the deposition processes taking place in both the gaseous and the solid phases involved is needed. The works published in this special number represent an excellent illustration of the recent achievements reached by means of the combination of fundamental/applied experimental research and numerical modelling, increasing this knowledge and enabling the progress in new tailored nanostructured surfaces and thin films. The potential that these developments suggest for the future of surface engineering can only be glimpsed nowadays.

Funding: This research was funded by the FEDER program through the Junta de Andalucía (PAIDI-2020) grant number P18-RT-3480, the University of Seville (V PPIT-US) and the Comunidad Autónoma de Madrid (IND2017/IND-7668).

Acknowledgments: Rafael Alvarez would like to thank all authors who submitted their research to this Special Issue, the referees who reviewed the submitted manuscripts, and the Assistant Editor, Mirabelle Wang, who made it all work.

Conflicts of Interest: The authors declare no conflict of interest.

References

1. Baptista, A.; Silva, F.J.G.; Porteiro, J.; Míguez, J.L.; Pinto, G. Sputtering Physical Vapour Deposition (PVD) Coatings: A Critical Review on Process Improvement and Market Trend Demands. *Coatings* **2018**, *8*, 402. [CrossRef]

2. Palmero, A.; Tomozeiu, N.; Vredenberg, A.; Arnoldbik, W.; Habraken, F. On the deposition process of silicon suboxides by a RF magnetron reactive sputtering in Ar–O_2 mixtures: theoretical and experimental approach. *Surf. Coat. Technol.* **2004**, *177*, 215–221. [CrossRef]
3. *Reactive Sputter Deposition*; Series in Materials Science; Springer Science and Business Media LLC: Berlin/Heidelberg, Germany, 2008; Volume 109.
4. Palmero, A. van Hattum, E.D.; Arnoldbik, W.M.; Vredenberg, A.M.; Habraken, F.H.P.M. Characterization of the plasma in a radio-frequency magnetron sputtering system. *J. Appl. Phys.* **2004**, *95*, 7611. [CrossRef]
5. Garcia-Valenzuela, A.; Alvarez, R.; Rico, V.; Cotrino, J.; Elipe, A.R.-G.; Palmero, A. Growth of nanocolumnar porous TiO_2 thin films by magnetron sputtering using particle collimators. *Surf. Coat. Technol.* **2018**, *343*, 172–177. [CrossRef]
6. Garcia-Valenzuela, A.; Muñoz-Piña, S.; Alcala, G.; Alvarez, R.; Lacroix, B.; Santos, A.J.; Cuevas-Maraver, J.; Rico, V.; Gago, R.; Vazquez, L.; et al. Growth of nanocolumnar thin films on patterned substrates at oblique angles. *Plasma Process. Polym.* **2019**, *16*, 1800135. [CrossRef]
7. Elmkhah, H.; Attarzadeh, F.; Fattah-Alhosseini, A.; Kim, K.H. Microstructural and electrochemical comparison between TiN coatings deposited through HIPIMS and DCMS techniques. *J. Alloy. Compd.* **2018**, *735*, 422–429. [CrossRef]
8. Elsheikh, A.H.; Sharshir, S.W.; Ali, M.K.A.; Shaibo, J.; Edreis, E.M.; Abdelhamid, T.; Du, C.; Zhang, H. Thin film technology for solar steam generation: A new dawn. *Sol. Energy* **2019**, *177*, 561–575. [CrossRef]
9. Zheng, X.; Zhang, X.Y.; Du, S.; Liu, J.; Yang, Z.; Pang, X. A Review on Design and Research Progress of Antifriction and Wear-Resistant Multilayer Coatings. *Mater. Rep.* **2019**, *33*, 444–453. [CrossRef]
10. Olayinka, A.; Akinlabi, E.; Oladijo, P. Influence of TiC thin film growth morphology deposited by RF magnetron sputtering on the mechanical and tribology properties of Ti_6Al_4V. *Mater. Today* **2020**, *26*, 1469–1472.
11. Ramezannezhad, M.; Nikfarjam, A.; Hajghassem, H.; Akram, M.M.; Gazmeh, M. A micro optofluidic system for toluene detection application. *Microelectron. Eng.* **2020**, *222*, 111204. [CrossRef]
12. Wang, S.; Kravchyk, K.V.; Filippin, A.N.; Müller, U.; Tiwari, A.N.; Buecheler, S.; Bodnarchuk, M.I.; Kovalenko, M.V. Aluminum Chloride-Graphite Batteries with Flexible Current Collectors Prepared from Earth-Abundant Elements. *Adv. Sci.* **2018**, *5*, 1700712. [CrossRef] [PubMed]
13. Colin, J.; Jamnig, A.; Furgeaud, C.; Michel, A.; Pliatsikas, N.; Sarakinos, K.; Abadias, G. In Situ and Real-Time Nanoscale Monitoring of Ultra-Thin Metal Film Growth Using Optical and Electrical Diagnostic Tools. *Nanomaterials* **2020**, *10*, 2225. [CrossRef] [PubMed]
14. Zhao, X.; Wen, J.; Zhu, A.; Cheng, M.; Zhu, Q.; Zhang, X.; Wang, Y.; Zhang, Y. Manipulation and Applications of Hotspots in Nanostructured Surfaces and Thin Films. *Nanomaterials* **2020**, *10*, 1667. [CrossRef] [PubMed]
15. Chargui, A.; El Beainou, R.; Mosset, A.; Euphrasie, S.; Potin, V.; Vairac, P.; Martin, N. Influence of Thickness and Sputtering Pressure on Electrical Resistivity and Elastic Wave Propagation in Oriented Columnar Tungsten Thin Films. *Nanomaterials* **2020**, *10*, 81. [CrossRef] [PubMed]
16. Stamate, E. Spatially Resolved Optoelectronic Properties of Al-Doped Zinc Oxide Thin Films Deposited by Radio-Frequency Magnetron Plasma Sputtering Without Substrate Heating. *Nanomaterials* **2019**, *10*, 14. [CrossRef]
17. Liang, Y.-C.; Liu, Y.-C. Design of Nanoscaled Surface Morphology of TiO_2–Ag_2O Composite Nanorods through Sputtering Decoration Process and Their Low-Concentration NO_2 Gas-Sensing Behaviors. *Nanomaterials* **2019**, *9*, 1150. [CrossRef]
18. Ma, L.; Wang, J.; Huang, H.; Zhang, Z.; Li, X.; Yi, F. Simultaneous Thermal Stability and Ultrahigh Sensitivity of Heterojunction SERS Substrates. *Nanomaterials* **2019**, *9*, 830. [CrossRef]
19. Panepinto, A.; Snyders, R. Recent Advances in the Development of Nano-Sculpted Films by Magnetron Sputtering for Energy-Related Applications. *Nanomaterials* **2020**, *10*, 2039. [CrossRef]
20. Alvarez, R.; Muñoz-Piña, S.; González, M.U.; Izquierdo-Barba, I.; Fernández, I.; Rico, V.; Arcos, D.; Garcia-Valenzuela, A.; Palmero, A.; Vallet-Regí, M.; et al. Antibacterial Nanostructured Ti Coatings by Magnetron Sputtering: From Laboratory Scales to Industrial Reactors. *Nanomaterials* **2019**, *9*, 1217. [CrossRef]
21. Jeong, H.-Y.; Nam, S.-H.; Park, K.-S.; Yoon, S.-Y.; Park, C.; Jang, J. Significant Performance and Stability Improvements of Low-Temperature IGZO TFTs by the Formation of In-F Nanoparticles on an SiO_2 Buffer Layer. *Nanomaterials* **2020**, *10*, 1165. [CrossRef]

22. Tseng, C.-C.; Wu, G.; Chang, L.-B.; Jeng, M.-J.; Feng, W.-S.; Chen, D.W.; Chen, L.-C.; Lee, K.-L. Effects of Annealing on Characteristics of $Cu_2ZnSnSe_4/CH_3NH_3PbI_3/ZnS/IZO$ Nanostructures for Enhanced Photovoltaic Solar Cells. *Nanomaterials* **2020**, *10*, 521. [CrossRef] [PubMed]

Publisher's Note: MDPI stays neutral with regard to jurisdictional claims in published maps and institutional affiliations.

© 2021 by the authors. Licensee MDPI, Basel, Switzerland. This article is an open access article distributed under the terms and conditions of the Creative Commons Attribution (CC BY) license (http://creativecommons.org/licenses/by/4.0/).

Article

Recent Advances in the Development of Nano-Sculpted Films by Magnetron Sputtering for Energy-Related Applications

Adriano Panepinto [1,*] and Rony Snyders [1,2]

1. Chemistry of Plasma-Surface Interactions, University of Mons, 23 Place du Parc, 7000 Mons, Belgium; rony.snyders@umons.ac.be
2. Materia Nova Research Center, Chemistry of Plasma-Surface Interactions, 3 Avenue Nicolas Copernic, 7000 Mons, Belgium
* Correspondence: adriano.panepinto@umons.ac.be; Tel.: +32-(65)-554945

Received: 30 July 2020; Accepted: 13 September 2020; Published: 15 October 2020

Abstract: In this paper, we overview the recent progress we made in the magnetron sputtering-based developments of nano-sculpted thin films intended for energy-related applications such as energy conversion. This paper summarizes our recent experimental work often supported by simulation and theoretical results. Specifically, the development of a new generation of nano-sculpted photo-anodes based on TiO_2 for application in dye-sensitized solar cells is discussed.

Keywords: magnetron sputtering; GLAD; nano-sculpted films; growth simulations; DSSCs

1. Introduction

In addition to more conventional features such as the thickness, chemical composition, or phase constitution, the control of the morphology at the submicrometer scale and the associated porosity is accepted as an essential parameter, allowing the properties of thin films to be tailored. Yet, for many years, the design and fabrication of nano-sculpted thin films have been recognized as a new opportunity to improve the performance of the thin films in a wide variety of applications ubiquitous in our society in the fields of microelectronics, information processing, as well as energy generation and storage. [1] The main interest toward nano-sculpted films mainly arises from their high surface-to-volume ratio, allowing for an important developed surface that can be used as examples to accommodate a large quantity of molecules (i.e., in dye-sensitized solar cells) or to strongly increase the active sites (i.e., in photocatalytic materials).

Specifically addressing the domain of solar energy harvesting, one of the most frequent nano-sculpted film architectures is based on fritted nanoparticles thin films [2]. Such a structure, which is the one used nowadays in dye-sensitized solar cells (DSSCs) as an example, often presents a very high specific surface area and very good porosity, which is crucial to improving the dye absorption in the mentioned application. Nevertheless, they also present a dramatic limit in terms of the relatively low quality of charge carriers transport [3,4]. This problem is often associated with the scattering of the charge carriers at the numerous grain boundaries between the fritted particles constituting the film [5]. This has motivated the development of thin films based on one-dimensional (1D) nanostructures, expected to present a much better facility to transport charge through the film. In reality, such a thin film consists of a tri-dimensional (3D) arrangement constituted of nano-objects grown with a preferential direction from the substrate. This means that two of their dimensions are less than 100 nm while the other can reach several micrometers [6]. Individually, such a 1D object would avoid the carriers' transport limitations by creating direct pathways through the material without particle interconnections, while keeping a high specific surface area. As a consequence, the transport path for the charge collection is efficiently shortened and, in most cases, these materials have higher electron diffusion coefficients than nonordered nanostructures [7], allowing a large increase in the thickness of the film. The price to pay for such an architecture is an often lower specific surface area, as well as porosity, in comparison with the conventional nanoparticles-based thin-film architecture. Indeed, for the latter, the conventional specific surface area and porosity are ~200 m^2/g and >75%, respectively, while the best reported values for 1D nanostructures-based films are ~140 m^2/g and 70%, respectively [8,9]. This has motivated many efforts to improve such a structure in the last few years. Basically, researchers have been focused on the improvement in the design of these hierarchical nanostructures, aiming to reach an optimal equilibrium between the specific surface area, porosity, and charge transport efficiency. As a noteworthy example, Kuang et al. established a systematical strategy to grow TiO_2 hierarchical nanostructures made of nanowires (NW) on which nanorods (NR) are branched, themselves being sources of nanorods (NR), labeled as NW/NR/NR. These hierarchical structures have been utilized in solar cell applications, allowing for a conversion efficiency of up to 9% due to a larger specific surface area, lower transport time, and longer electron lifetime than TiO_2 nanoparticles [10]. From these results, they concluded that the design of 1D hierarchical nanostructure thin films is one of the keys to find the best agreement between a high carriers' reservoir capability and good charge transport properties in DSSCs.

Numerous processes have been used to synthesize such nano-sculpted materials: Anodic oxidation [11], electron beam evaporation [12], atomic layer deposition [13,14], sol–gel deposition in a template [15,16], or hydrothermal methods, which has been intensively utilized by successive treatments to obtain branched nanowires [17,18]. However, these methods are usually difficult to industrialize, often lead to the synthesis of amorphous materials, and make necessary the use of solvents and toxic chemicals. In this context, it is necessary to develop industrially viable alternative synthesis roads for such structures that would allow a crystallized material to be synthesized with a low environmental impact. Physical vapor deposition (PVD) techniques are well-established in various manufacturing areas such as microelectronics, automotive, and biomedical industries [19]. In these fields, plasma-assisted processes are generally preferred to thermal evaporation in response to requirements of materials processing at reasonable temperatures. In PVD processes, the target material to be deposited as a thin film is transformed into the vapor phase by different means, generally involving plasma generation (except for thermal evaporation). The chemical composition of the deposited film can be tuned by the addition of various reactive gases (O_2, N_2, etc.) in order to form oxides, nitrides, or more complex compounds, which makes the technique versatile.

As a widespread plasma technology, magnetron sputtering, which belongs to the PVD methods and consists of bombarding a target material with accelerated ions from the plasma, leading to the ejection of particles (mainly atoms and clusters), has been used to build the thin film. Oxides, nitrides, or carbides can also be grown by adding pure O_2, N_2, or C-based vapor sources inside the magnetron sputtering deposition chamber, the so-called reactive magnetron sputtering (RMS) regime [20,21]. The magnetron sputtering process offers the opportunity for tuning the crystalline constitution depending on the energy brought to the growing film by adjusting the experimental parameters such as the applied power [22]. In most of the aforementioned applications, magnetron-deposited films are meaningful because they are dense, homogeneous, and chemically pure [23].

In this work, in order to generate a 1D structure-based thin film, the so-called nano-sculpted films, we have utilized magnetron sputtering in glancing angle geometry. Glancing angle deposition (GLAD) is a particular case of oblique deposition where the substrate position is manipulated during the film deposition [24]. The technique takes advantage of the ballistic shadowing effect, which allows the formation of columnar microstructures as the film is growing. The basic operation principle is presented in Figure 1 and can be summarized as follows: While the substrate is tilted with an angle α compared to the target normal, the initial nuclei of the depositing film randomly roughen the surface. Subsequently, the depositing particles nucleate on the substrate, while the region behind the nucleus does not receive any vapor, because it falls in the shadow of the nucleus. Consequently, a larger number of particles will be deposited onto the nuclei than in the shadowed area. This inequality increases as growth continues. As only the tops of the nuclei receive the depositing material, the nuclei will develop into columns, tilted in the direction of the incident particles flux and forming an angle β with the substrate normal ($\beta < \alpha$). The β value depends on many experimental parameters as it will be discussed in this paper.

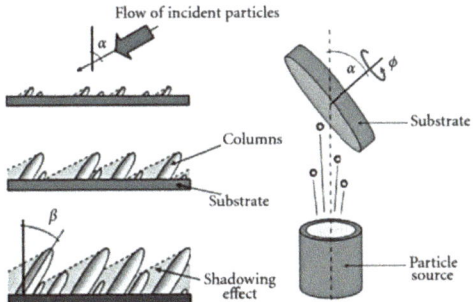

Figure 1. Schematic description of the ballistic shadowing effect during thin-film growth in glancing angle geometry (reproduced from [25], Hindawi, 2012).

The key principle of GLAD consists of changing the vapor flux direction to operate ballistic shadowing and to provide control over the final thin-film morphology. Two degrees of freedom are obtained by tilting the substrate with respect to the source of particles (α), while rotating the substrate around its normal axis allows the substrate azimuthal angle ϕ to be controlled with a fixed substrate rotation speed ($\dot{\phi}_s$). Varying ϕ modifies the direction of the incident vapor flux and provides control over the shadowed regions of the substrate.

There are mainly four archetypal columnar microstructures, which illustrate how substrate motion affects the microstructure. Basically, inclined columns are obtained when working with a fixed tilt angle (α) higher than 60°; zig-zags are generated when rotating the substrate by successive rotation of the substrate by a 180° angle in the latter configuration; and plots or helical structures can be grown continuously, rotating the tilted substrate during deposition [9]. All of these structures are generated by modifying the substrate rotation as α is unchanged for each one. The ability to sculpt the film and access the various morphologies is provided by the trajectory of the incident vapor flux relative to the substrate surface during deposition. Accordingly, each nanostructure is characterized by a given porosity (inter-columnar space) that mainly depends on the columnar tilt.

It is important that the incident vapor remains highly directional to avoid merging of the shadowed regions. Indeed, deposition with a poorly collimated incident flux allows vapor to directly access the shadowed area. This implies that the mean free path of the particles should be greater than the distance to the substrate. It is thus evident that PVD techniques such as thermal and electron beam evaporation are the most prevalent in GLAD research because they allow high target-to-substrate distances and low operating pressures with small vapor sources. Therefore, GLAD has mainly been utilized in combination with an evaporation source to grow various nano-sculpted materials [26]. Nevertheless, using evaporation, the energy brought to the growing film does not allow us to control the crystalline structure of the deposited films in most of the cases. By contrast, as mentioned, magnetron sputtering is a recognizable technique to do this even without intentional heating of the growing material. Indeed, in this case, the crystallization is promoted by the bombardment of energetic particles, as well as infrared radiation emitted from the target during the sputtering process [27]. Although promising, the utilization of GLAD geometry in combination with magnetron sputtering, i.e. magnetron sputtering in glancing angle geometry (MS-GLAD), is surprisingly quite recent [28], and has attracted considerable interests for 10 years [29].

In this paper, we aim to overview the work that has been developed in our group during the past few years, utilizing this original synthesis process to design nano-sculpted thin films that are ultimately utilized in energy-related applications. Due to the unusual character of the MS-GLAD, we first had to answer many questions related to the growth mechanism of the nano-sculpted films by this approach in order to be able to control the synthesized films features (morphology, chemistry, crystalline constitution, etc.). Indeed, if magnetron-sputtered thin films can be crystallized by a higher supply of energy, the price to pay is the generation of dense films, inhibiting the effect of the grazing mode configuration [30]. In order to obtain a full picture of the growth mechanism, experimental as well as simulation works have been developed. Finally, we will summarize the benefits that can be associated with the utilization of these nano-sculpted films in energy-related applications, specifically as the photo-anode in dye-sensitized solar cells (DSSCs), although other applications of our films have been investigated in our group [31].

2. Materials and Methods

2.1. Experimental Setup

The nano-sculpted films were synthesized in a cylindrical stainless-steel magnetron sputtering chamber (height: 60 cm, diameter: 42 cm), schematically presented in Figure 2. The chamber was evacuated down to a residual pressure of 10^{-4} Pa by a turbo molecular pump (Edwards nEXT400D 160W, Edwards, Irvine, California, CA, USA), backed by a dry primary pump (Edwards nXDS10i, Edwards, Irvine, California, CA, USA).

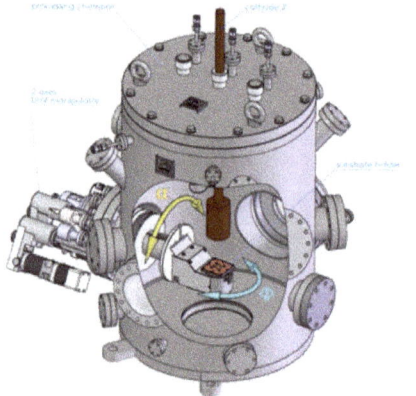

Figure 2. Sketch of the deposition chamber used in this work (reproduced from [32], MDPI, 2019).

An unbalanced magnetron cathode was installed in front of the substrate, at the top of the chamber on which a 2 in. (5.08 cm) diameter and 0.25 in. (0.635 cm) thick target was connected. The target/substrate distance was fixed at 7 cm. Pure Ti and Mg (both with 99.99% purity) were used as target materials in this study. In order to modify the phase constitution of the deposited films, the target was sputtered either in direct current (DC) mode or in the high power impulse magnetron sputtering (HiPIMS) regime. For the DC mode, an Advanced Energy MDK 1.5 K (Advanced Energy, Denver, Colorado, CO, USA) power supply was used. The power (P) was fixed at 150 W, corresponding to a power density on the target surface of 7.5 W·cm^{-2}, which was calculated by taking into account the target surface exposed to the plasma (~20 cm^2). In the HiPIMS regime, a lab-made power supply based on a one-quadrant chopper topology was used, allowing the generation of short high-power pulses at the cathode [33]. We have to mention that the discharge voltage is measured at the output of the lab-made power supply, not directly at the cathode. More information can be found elsewhere [34].

Argon, which is the sputtering gas with or without O_2 as the reactive gas (both with 99.999% purity), was introduced in the chamber using two distinct mass flow meters in order to grow oxide or metallic compound, respectively. Note that the gases were mixed prior to being injected in the vacuum chamber. All thin films were deposited at constant total gas flux. It was fixed at 15 sccm (standard cubic centimeter per minute) to allow low working pressure (0.13 Pa) according to the pumping rate.

The substrate was installed on a 2-axis manipulator, allowing two rotation motions: Along the α angle to tilt the substrate from $\alpha = 0°$ to $\alpha = 90°$ with regard to the cathode axis and/or along the ϕ angle to rotate the substrate step by step or in continuous mode with a given angular speed $\dot{\phi}_s$. The α and ϕ angles were varied in order to generate various morphologies: Discrete rotations ($\phi = +180°$ or $-180°$) allow zigzag structures to be grown, while continuous rotations ($\dot{\phi}_s = 0.1, 1.0$ or $10°/s$) lead to vertical pillars and helicoidal structures.

Silicon single crystals with an (100) orientation and whose resistivity is $5 \cdot 10^{-3}$ $\Omega.cm$, or fluorine-doped tin oxide (FTO)-coated glasses, were utilized as substrates depending on the subsequent type of characterization. The substrates were cleaned with detergent solution, rinsed with ultra-pure water, and placed at the ground potential and at ambient temperature prior to deposition.

2.2. Characterization Techniques

Field emission gun scanning electron microscopy (FEG-SEM Hitachi SU8020, Hitachi, Tokyo, Japan) was used to observe the microstructure of the nanostructured films, while the nanostructure was investigated by transmission electron microscopy (TEM Philips CM200, Philips, Amsterdam, Netherlands). The cross-sectional lamellae of the untreated nanostructured films were prepared by mechanical polishing and ion milling. Individual columns of the single crystalline thin film were scratched to observe each column separately.

Grazing incidence X-ray diffraction (GIXRD) analysis (Panalytical Empyrean, Malvern Panalytical, Malvern, UK) was used to determine the phase constitution of the samples. The Cu Kα_1 source (1.5406 Å) was used and the X-ray source voltage was fixed at 45 kV and a current at 40 mA.

The experimental procedure used to design the DSSCs is described in depth elsewhere [35]. Briefly, the dye grafting of TiO_2-based nano-sculpted electrodes (0.25 cm^2) was performed by immersion overnight in a solution of acetonitrile and tertbutyl alcohol (volume ratio: 1/1) containing dye sensitizer (0.3 mmol) and (3R),(7R)-dihydroxy-5-cholic acid (Sigma-Aldrich, Saint-Louis, Missouri, MO, USA) (2 mmol) to avoid aggregation of the dye. Then, the sensitized electrode was assembled with a platinized FTO electrode, both separated by 25 µm-thick Surlyn® (Dow Chemical, Midland, Michigan, MI, USA) to prevent the electrolyte from leaking. The internal space was filled with a liquid electrolyte by using a vacuum backfilling system. The photovoltaic performances of the cells were then measured under a simulated air mass (AM 1.5) Global spectrum and 1000 W/m^2 illumination.

2.3. Simulation Protocol

The simulation of nano-sculpted film growth is possible with kinetic Monte Carlo (kMC) algorithms [36,37]. This approach is useful for the modeling of various surface processes such as the nucleation, growth, structural modifications, or dynamic evolution of obliquely deposited structures [38,39]. These features are implemented into the freely distributed NASCAM (nanoscale modeling) code (version 4.6.2 rev. 6; University of Namur: Namur, 2018) [40,41], which is particularly suitable to simulate GLAD processes as it takes into account the motion of the substrate during deposition (translation, rotation, and oscillation).

In this code, the incoming vapor flux is represented by hard spheres and their mobility is simulated according to the ballistic deposition approximation for minimizing the computation time. Taking into account the energy and the angular distribution of the vapor source, those atoms travel toward the substrate along linear trajectories. Then, the deposited particles become part of the growing film.

As input, the code uses the kinetic energy and angular distribution of the sputtered atoms calculated by SRIM (stopping and range of ions in matter, SRIM-2013; Chester, Maryland, MD, 21619,

USA) [42] and SIMTRA (simulation of the metal transport, version 2.2; University of Ghent: Ghent, 2018) codes, respectively. Indeed, the energy and the direction of the particles that are sputtered from the target material are first calculated by SRIM. Then, SIMTRA simulates the transport of these species toward the substrate, taking into account all collisions happening in the gas phase. At the end, the PoreSTAT plugin can be used to evaluate the porosity of the simulated films from the NASCAM output files [43]. This simulation strategy is presented in Figure 3.

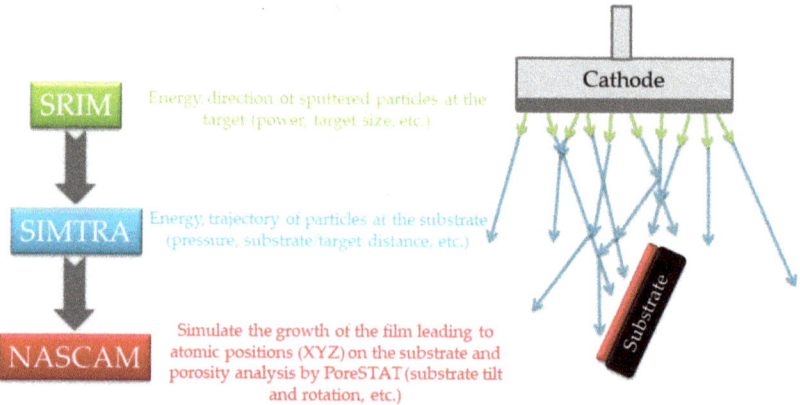

Figure 3. Kinetic Monte Carlo simulation strategy.

The mobility of the atoms that reach the substrate is severely dependent on the energy of incoming deposited atoms. However, the morphology of the films depends on the different deposition parameters [44–47]. Consequently, to simulate the growth of thin films synthesized at high temperature or with high-kinetic-energy incoming atoms, the approximation of ballistic deposition has to be completed by the diffusion phenomenon [22,27]. Diffusion and evaporation events can take place between two atom depositions at an equal time interval determined by the deposition rate. For each step of the simulation, a list of atoms that can diffuse at the surface or evaporate is created for each possible physical event. The evolution of the system is, thus, determined by the probabilities of the events that may occur during the simulation. This probability can be implemented into NASCAM code via their activation energies (E_a), which can either be found in the literature or calculated by molecular dynamics or potential models.

For each working condition, the energy and the angular distribution of the species can be adapted by the introduction of the experimental parameters such as the working pressure, the power applied to the target, the racetrack size, and the target-to-substrate distance. In order to compare simulated and experimental thin films having the same thickness, the number of deposited atoms (N) and the substrate size (XYZ) can be tuned in the NASCAM input file. X and Y correspond to the length and width of the substrate, respectively, while Z accounts for the height of the deposited film in atom units.

2D NASCAM simulations were performed for direct comparison with the cross-sectional film morphology, while 3D simulations were performed for the porosity evaluation. The Ti and Mg deposition rates were fixed at 0.5 monolayers per second (0.16 and 0.30 nm/s, respectively), which is of the same order of magnitude in comparison with their experimental values (0.17 and 0.32 nm/s, respectively).

3. Results and Discussion

In this section, we overview part of our recent works on the synthesis of nano-sculpted thin films. First, our understanding of the growth mechanisms of these nano-sculpted materials is presented. Then, their utilization in DSSC applications is demonstrated and discussed.

3.1. Growth Mechanisms of Metallic Films

In the first attempt, we establish the deposition parameters, allowing the growth of nano-sculpted metallic thin films by MS-GLAD. Metallic materials were chosen because of their easiest implementation in the NASCAM software. Ti and Mg were chosen as model materials because their growth mechanism is expected to be different according to the structure zone model (SZM) of vacuum-deposited materials published by Movchan and Demchishin [48] (see Figure 4) This model describes the morphologies of the deposited films according to the homologous temperature T_s/T_m, where T_s is the substrate temperature and T_m is the melting point of the deposited material. Three zones characterizing different film morphologies, mainly depending on surface diffusion phenomena, are defined as a function of T_s/T_m. Zone 1, corresponding to $T_s/T_m < 0.3$, is characterized by a low diffusion and mobility of the adsorbed atoms, leading to a columnar structure with fibrous morphology and weakly bounded grains. For $0.3 < T_s/T_m < 0.5$, Zone 2 is reached and the surface diffusion of the adatoms has a dominant effect on the film growth, leading to a tightly packed columnar grain structure. Finally, Zone 3, obtained for $T_s/T_m > 0.5$, corresponds to the condition for which bulk diffusion processes are activated, enabling the diffusion of the grain boundary, as well as the recrystallization during the film growth. For room-temperature growth conditions, T_s/T_m is 0.15 and 0.32 for Ti and Mg, respectively, which clearly highlights different surface diffusion behaviors during the growth of these materials.

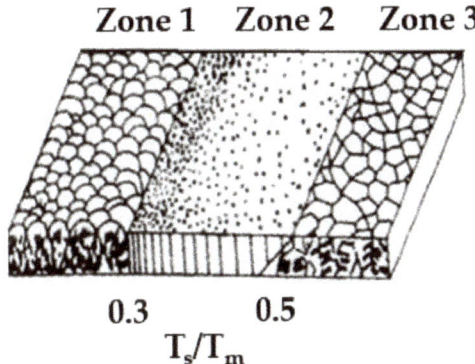

Figure 4. Structure zone diagram of Movchan and Demchishin applicable to the film growth by physical vapor deposition (PVD) (reproduced from [48] with permission from Springer, 2014).

Furthermore, from an application point of view, TiO$_2$ is of great interest as the charge transport layer in DSSCs [2], while Mg and its hydride are promising candidates for hydrogen storage [49].

3.1.1. Effect of the Angle of Deposition

The effect of the angle of deposition, α, on the morphological properties of Ti and Mg thin films was first investigated. For this set of experiments, the Ti and Mg films were deposited using high power (150 and 50 W, respectively) and low pressure (0.13 and 0.23 Pa, respectively) in order to generate a ballistic flux by maximizing the deposition rate (~10 and ~19 nm/min, respectively). The thickness of the deposited films was about (500 ± 50) and (620 ± 20) nm for Ti and Mg films, respectively. Ti and Mg thin films synthesized for various α angles from normal incidence ($\alpha = 0°$) to grazing angle ($\alpha = 89°$) with the corresponding kMC simulations are shown in Supplementary Materials, Figures S1 and S2. A number of 1.2×10^5 and 5×10^5 atoms was chosen for similar experimental and simulated film thicknesses according to the size of the Ti ($X = 250$ and $Y = 2$ Ti atom units) and Mg ($X = 1000$ and $Y = 2$ Mg atom units) simulation boxes, respectively. The evolution of the morphological properties assessed by the analysis of the cross-sectional SEM (scanning electron microscopy) images, as well as by the NASCAM simulation tool for both Ti and Mg nano-sculpted materials as a function of α, is presented in Figure 5.

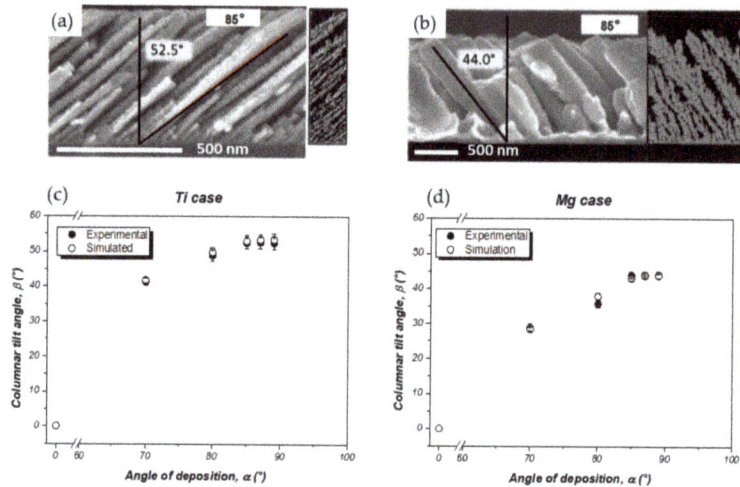

Figure 5. SEM (scanning electron microscopy) cross-sectional view with the corresponding simulation of: (**a**) Ti and (**b**) Mg columnar films deposited with $\alpha = 85°$. Evolution of the columnar tilt angle (β) as a function of the angle of deposition (α) for (**c**) Ti and (**d**) Mg nano-sculpted thin films. The error bars were estimated by taking the tilt average over 20 columns (adapted from [50] with permission from Elsevier, 2017, and from [32], MDPI, 2019).

As expected, the increase in α leads to generation of a nano-sculpted thin film composed by well-separated tilted columns in both cases (Ti and Mg materials). As explained in the introduction part, the columnar structure is strongly influenced by shadowing effects, particularly at extreme oblique incidence angles (>60°). If many features are common between both systems, we can notice that the width of the columns is higher for Mg than for Ti (see Figure 5a,b). This is explained by the higher homologous temperature for the Mg deposition; ~0.35 vs. ~0.19 for Ti, which means that the growth of Ti thin films belongs to the Zone 1 regime, while the growth of Mg thin films belongs to the Zone 2 regime of the SZM, allowing surface diffusion of the Mg adatoms, which ultimately leads to less dense and thicker columns compared to the Ti case.

Figure 5c,d shows that, for both metals, the columnar tilt angle (β) and the inter-columnar space drastically increase in these conditions. However, the β value stabilizes for $\alpha \geq 85°$, whatever the considered metal. This observation can be explained by the specific geometry of our experimental setup. Indeed, the diameter of the target (5 cm) has to be taken into account to distinguish α from the incident angle of the particles as the substrate-to-target distance (8 cm) is not very high in our geometry. Hence, the majority of the deposited particles comes from the racetrack region of the target. The size of the particles source thus induces a deviation in the α or ϕ directions corresponding to the angular distribution of the particles reaching the substrate, and increases with the target diameter. The width of the sputter flux distribution at 0.13 Pa, considering the geometry we used, was determined by using SIMTRA, and the angle of deviation was simulated to be around 10° (see Supplementary Materials, Figure S3). A stabilization of the β value for $\alpha \geq 85°$ was observed because the geometrical inclination of the substrate leads to an asymmetric deposition, and the particles sputtered at the left side of the target have a higher probability to reach the left side of the substrate compared to those sputtered on the right side.

All these behaviors are in almost perfect agreement with the simulated data, as shown in Figure 5c,d, which demonstrate the relevance of the modeling approach that is utilized. From these data, it is finally possible to highlight the importance of surface diffusion in the generation of such nano-sculpted metallic films.

On the other hand, a systematic lower β value in comparison with α is observed in all situations for both metals. This might be understood by the parallel momentum (kinetic energy) conservation principle [51]: under oblique incidence, the amount of kinetic energy conserved in the direction parallel to the film surface is determined by the angle of incidence. Thus, β varies with α and the kinetic energy of the impinging vapor atoms.

Based on these initial works, $\alpha = 85°$ was chosen for all subsequent thin-film syntheses as it allows well-defined columnar structures to be produced at reasonable deposition rates.

3.1.2. Effect of the Deposition Pressure

The deposition pressure (P_{dep}) is another important parameter in GLAD experiments. In this part, P_{dep} was varied from 0.13 to 1.3 Pa, while the sputtering power was fixed at 50 W for Mg and 150 W for Ti, and α was fixed at 85° in both cases. As this parameter does not influence the surface diffusion of the adatom, a similar dependence is observed for Ti and Mg. Therefore, we will only discuss the results obtained for the Ti films. A detailed description of the Mg case can be found elsewhere [32], and the cross-sectional SEM images associated with the simulated data are presented in the Supplementary Materials, Figure S4.

From the data shown in Figure 6, we learn that β rapidly decreases as P_{dep} increases, from $(52.7 \pm 2.0)°$ for 0.13 Pa to $(13.8 \pm 1.9)°$ for 1.3 Pa. As the collision probability increases with P_{dep}, this is attributed to a decrease in the collimation of the incident particle flux. Indeed, this probability is expressed by the mean free path of the sputtered atoms (λ) and is inversely proportional to P_{dep} as follows:

$$\lambda = \frac{k_B T}{\sqrt{2}\pi d^2 P_{dep}} \quad (1)$$

where k_B is the Boltzmann constant, T is the temperature in K, P_{dep} is the pressure in Pa, and d is the diameter of the gas particles in m [52]. Considering the atomic diameter of Ti particles (1.4 Å), λ ranges from 45 to 4.5 cm between 0.13 and 1.3 Pa, respectively. Owing to the target-to-substrate distance used in this work (8 cm), an increase in P_{dep} to 1.3 Pa induces a large amount of collisions between particles, resulting in a divergent particle flux reaching the substrate, and the film becomes denser.

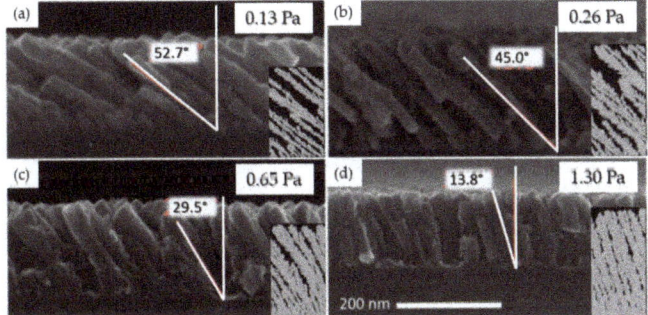

Figure 6. Ti thin films synthesized at 150 W and 85° for various deposition pressures: (**a**) 0.13; (**b**) 0.26; (**c**) 0.65; (**d**) 1.3 Pa with the corresponding simulations. The red angle accounts for the average columnar tilt angle (β) estimated by over 20 columns (adapted from [50] with permission from Elsevier, 2017).

The comparison between experimental and calculated values of β as a function of P_{dep}, as well as the calculated λ value for Ti atoms, is shown in Figure 7. The critical P_{dep} value from which λ becomes smaller than the target-to-substrate distance is estimated as ~0.7. Very few collisions occur through the vapor phase below this value, while numerous collisions between particles occur at higher pressure.

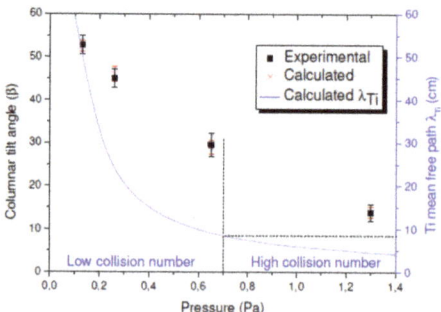

Figure 7. Columnar tilt angle as a function of the deposition pressure for experimental and simulated thin films. The blue line corresponds to the evolution of the mean free path of the sputtered Ti atoms calculated from Equation (1) (reproduced from [50] with permission from Elsevier, 2017).

SIMTRA calculations support this hypothesis as the simulated number of collisions between particles increases from 0.7 to 25 when increasing P_{dep} from 0.13 to 1.3 Pa. Furthermore, the angular distribution of the incoming Ti atoms has a low standard deviation (<30°) at 0.13 Pa, while it presents a large deviation (>100°) at 1.3 Pa [50].

Based on these analyses, the range of pressures where a ballistic deposition process occurs is, in our situations, from ~0.13 and ~0.26 Pa.

3.1.3. Effect of the Substrate Temperature

The substrate temperature (T_s) is one of the most important parameters influencing the growth mechanism of MS-GLAD thin films as the mobility of the adatoms, which is demonstrated to be a key element, can be activated through many diffusion phenomena by increasing T_s.

Tilted columnar Ti thin films were synthesized for 373 K < T_s < 873 K, where 373 K corresponds to the substrate surface temperature without intentional heating. This value is higher than the ambient temperature because of energy transfer phenomena occurring at the plasma-growing film interface associated with the IR radiation emanating from the Ti target [27]. The sputtering power was fixed at 150 W while P_{dep} = 0.13 Pa and α = 85°. Figure 8 shows that by increasing T_s, β decreases from (52.7 ± 1.8)° to (41.2 ± 1.5)°. This can be understood according to the SZM diagram discussed in Section 3.1 (see Figure 4). Indeed, based on this diagram, we learn that the films synthesized for T_s up to 523 K belongs to Zone 1 ($T_s/T_m \leq 0.27$), meaning that the surface diffusion is limited, allowing for the formation of porous and well-defined columnar morphologies. The geometric configuration governs the formation of microstructures in these conditions, leading to a strongly anisotropic deposition with a low influence of T_s on β. However, for 723 K < T_s < 873 K (i.e., 0.37 < T_s/T_m < 0.45), the films belong to Zone 2 of the SZM diagram. In that case, the anisotropic character of the deposited films is induced by the activation of surface diffusion, leading to a decrease in β.

In order to simulate the growth of nano-sculpted films at high temperature, the approximation of ballistic deposition, therefore, has to be completed by diffusion phenomena. The physical mechanisms that can be thermally activated, such as diffusion on a substrate and hopping from a substrate on an island, behave according to the following exponential law:

$$w_i = w_0 \exp\left(\frac{-\Delta E_a}{k_B T}\right) \quad (2)$$

where w_0 is the attempt rate that can be estimated as $w_0 = 2k_B T$, and ΔE_a is the activation energy for the physical mechanism.

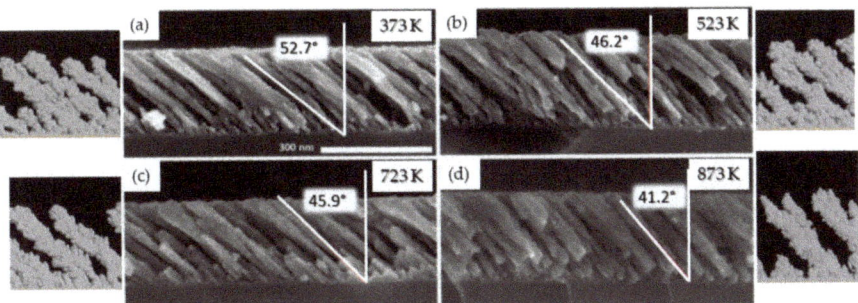

Figure 8. Ti thin films synthesized at 150 W, 0.13 Pa, and 85° for various substrate temperatures: (**a**) 373; (**b**) 523; (**c**) 723; (**d**) 873 K with the corresponding simulations. The red angle accounts for the average columnar tilt angle (β) estimated by over 20 columns (adapted from [50] with permission from Elsevier, 2017).

Based on many sets of simulations, it appears that the critical events of diffusion that affect the morphology of the films synthesized by GLAD are the hop up and hop down from one atomic layer to another ($E_{a,up}$ and $E_{a,down}$, respectively). The values leading to a better agreement with the experimental data are 2.0 and 2.5 eV for $E_{a,up}$ and $E_{a,down}$, respectively, while $E_{a,diffusion}$ is fixed at 1.35 eV. Finally, the low Ti vapor pressure (10^{-4} Pa at 1783 K [53]) strongly limits the probability that an evaporation event occurs in the temperature and pressure conditions used in this work, and is lower than a diffusion event. Consequently, $E_{a,evap}$ was fixed at 5.0 eV.

Simulations were, thus, performed for the same conditions than for the experiments to evaluate the impact of the diffusion phenomena on the film morphology (see Figure 8). The agreement between the experimental and simulated data validate that the diffusion is the reason for the evolution of β with T_s. Based on that, it is possible to adapt the empirical model proposed by Movchan and Demchishin for thin films deposited at oblique angles with different T_s [50].

Similar experiments were performed for the Mg case by varying T_s between 313 and 573 K with the other experimental conditions as follows: P = 50 W, P_{dep} = 0.26 Pa, and α = 85°. As previously discussed, Mg is a high-mobility material as its melting point is low enough to allow a Zone 2 growth mechanism at room temperature. The cross-sectional SEM images are presented in Figure 9.

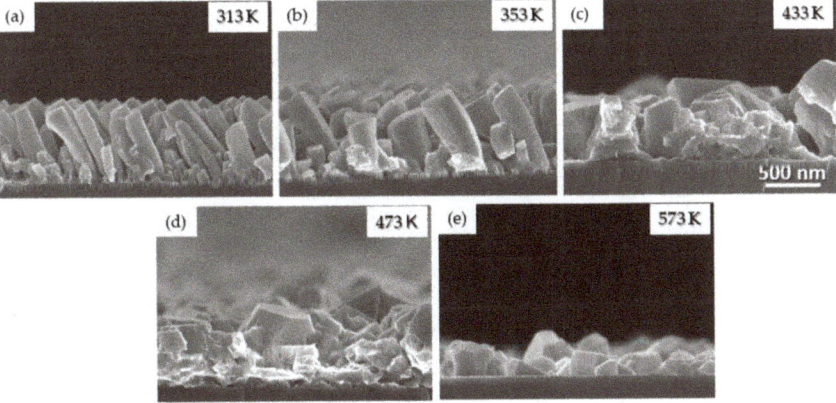

Figure 9. Mg thin films synthesized at 50 W, 0.26 Pa, and 85° for various substrate temperatures: (**a**) 313; (**b**) 353; (**c**) 433; (**d**) 473; (**e**) 573 K.

The films grown for T_s = 313 and 353 K, corresponding to a homologous temperature of ~0.34 and ~0.38, respectively, reveal a columnar structure with a decrease in β and a larger columnar width when increasing T_s, which is a similar behavior to Ti. For higher temperature, the columnar structure is lost and faceted grains appear. This is due to the activation of the bulk diffusion processes leading to bigger and recrystallized grains, as predicted by the SZM diagram in the Zone 3 regime that is reached in these conditions (T_s/T_m ~ 0.51 and 0.62, respectively). Figure 9c represents the intermediate situation (T_s/T_m ~ 0.47) where a weak columnar structure with large grains and reduced inter-columnar spaces can still be observed.

3.1.4. Effect of the Substrate Rotation

Using the optimal conditions established above (low T_s and P_{dep}, namely 373 K and 0.13 Pa, respectively), the effect of the rotation angle (ϕ) on the microstructure of the Ti film was evaluated. As a reminder, a variation in ϕ (+180° or −180°) leads to zigzag structures with various numbers of branches depending of the number of cycles, while vertical or helicoidal structures are generated for a continuous rotation of the substrate (0.1, 1.0, or 10.0°/s), as shown in Figure 10.

At the bottom of the images, we can observe the initial stages of the growth including a large number of small objects. Then, the number of nano-objects rapidly decreases due to the competitive growth occurring, while the film becomes thicker. This effect is more pronounced by the modification of substrate orientation during the generation of zigzag structures, which also induces a variation in β (see Figure 10a). The rotation of the substrate with ϕ modifies the local deposition geometry. The first nuclei are deposited with a uniform α onto a flat substrate, allowing a corresponding β. After that, the substrate is rotated at 180° and the incident vapor thus meets a surface with a different orientation because deposition then takes place on top of the first object. The effective local deposition angle α' that occurs during the deposition with α onto a tilted column with β is then $\alpha' = \alpha + \beta - 90°$ [54].

The corresponding simulated data are obtained without taking into account diffusion phenomena. The evolution of the growth, as well as the final microstructure (β variation between two zigzags), are reproduced well. The inter-column competition, explaining the decrease in the number of nano-objects during the film growth, starts during the nucleation stage. Hence, the evolution of the columnar growth is driven by this competition mechanism and generates a film morphology scale-invariant, in view of the different stages of the growth [55].

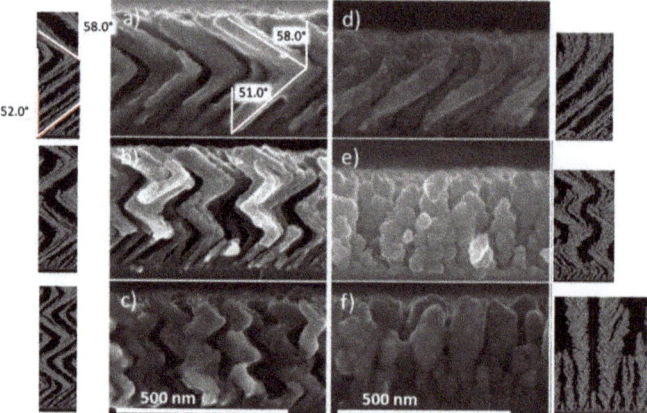

Figure 10. Ti thin films (150 W, 0.13 Pa, 85°) deposited at various rotations of the substrate to generate: (**a**–**c**) Zigzag structures; (**d**) helicoidal structures at 0.1°/s; vertical pillars at (**e**) 1.0°/s and (**f**) 10.0°/s with the corresponding simulations. The red angle accounts for the average columnar tilt angle (β) (adapted from [50] with permission from Elsevier, 2017).

3.1.5. Evolution of the Porosity

In order to obtain an evaluation of the porosity of our films, a NASCAM simulation has been used. This is one of the most important outputs of these simulations as it is very difficult to experimentally measure such a kind of porosity in thin-film materials. To do this, the output data of the NASCAM simulations were treated using the PoreSTAT plugin, which basically defines the size of the pores in order to evaluate the porosity (Φ) of the generated structures. In the case of Mg thin films, we observed a linear evolution of Φ with the so-defined aspect ratio (Γ), which corresponds to the ratio of the inter-columnar spaces on the width of the columns (see Figure 11), assessed from electron microscopy images. This parameter gives an indication of the specific surface area of the material. For example, more space between the columns or a reduction in the column width will lead to an increase in the material porosity, and thus, of Γ. Therefore, the tuning of the key parameters during the growth such as P_{dep} and α allows the film porosity to be finely monitored, which is very interesting in view of the applications that are foreseen.

For Ti thin films, similar simulations have been performed with even more details. Indeed, in this case, the porosity has been computed as an "effective porosity," Φ_e, because the porosity value depends on the size of the object that could penetrate the system. To do this, two molecules (M1 and M2) having different diameters (0.64 and 3.20 nm, respectively) have been considered to penetrate the Ti nano-sculpted films. The calculated ratio between the total porosity volume and the accessible porosity for M1 and M2 molecules penetration gives the Φ_e for each molecule. This gives a flavor of the organization of the porosity at the nanoscale. Many models have been performed for different types of Ti nano-sculpted films. Figure 12 summarizes the evolution of Φ_e with various experimental parameters for M1 and M2 molecules.

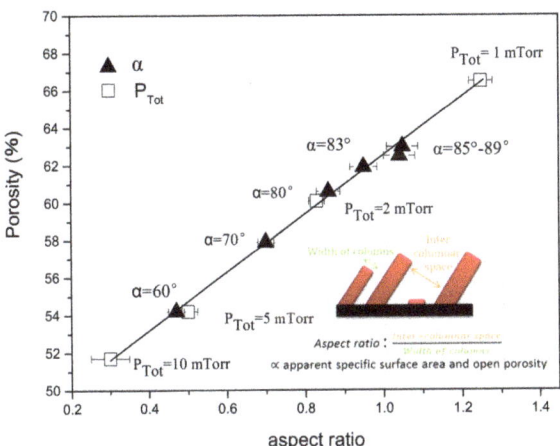

Figure 11. Evolution of the porosity as a function of the aspect ratio for Mg nano-sculpted films deposited at various deposition pressures and α. The inset illustrates the definition of the aspect ratio, Γ (adapted from [32], MDPI, 2019).

A significantly higher Φ_e value is obtained for the smallest molecule in all studied conditions: Around 60% for M1 vs. less over 5% for M2. This can only be explicated by the nanoscale structure of the film, suggesting a hierarchical growth, as observed in both TEM image and 2D simulation (see Figure 13): large micro-columns are formed by the agglomeration of the nano-columns upon increase in the film thickness. This is especially the case in low-mobility deposition conditions. Since then, the open-porosity from the nano-columns has not been available for the grafting of large molecules such as M2.

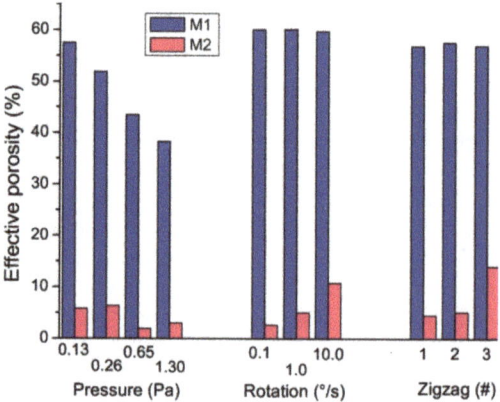

Figure 12. Kinetic Monte Carlo (kMC) tri-dimensional (3D) analyses of the effective porosity for different conditions with pore sizes above 0.64 nm (M1) and above 3.2 nm (M2) (reproduced from [50] with permission from Elsevier, 2017).

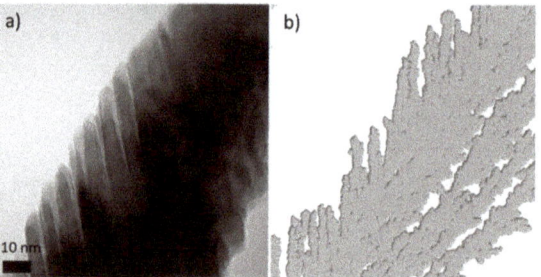

Figure 13. (a) Transmission electron microscopy (TEM) image of a Ti micro-column constituted by nano-columns from a Ti thin film synthesized at 0.13 Pa, 150 W, and 85°; (b) the corresponding kMC simulation where each atom is represented by its covalent radius (reproduced from [50] with permission from Elsevier, 2017).

The results presented in Figure 12 also revealed that: (i) high P_{dep} induces low Φ_e values, which is associated with the already discussed densification of the film, while (ii) the largest Φ_e values are obtained with a high substrate rotation speed or high number of zigzags. The latter behavior can be understood by the fact that the inter-columnar spaces become larger when a reduced number of growing columns are promoted because of the modification of the substrate orientation during deposition. Furthermore, the Φ_e associated with the M1 molecule is quite constant because of its small size. Indeed, M1 might impregnate the film even without variation in ϕ. Consequently, increasing the inter-columnar space by modifying ϕ does not influence the Φ_e value of sufficiently small molecules.

3.2. Nano-Sculpted Oxide Films

From our work on metallic films, we learn that the melting temperature and the associated reduced temperature is a key element, allowing the structure of nano-sculpted films synthesized by MS-GLAD to be understood. Therefore, we expected a different impact for TiO_2 and MgO thin films compared to their respective metallic form as the melting point of TiO_2 is close to that of Ti (~2116 vs. ~1941 K, respectively), while for MgO, the difference is important (~3125 vs. ~923 K for Mg). For both Ti and Mg cases, therefore, we synthesized the corresponding oxides by adding O_2 to the RMS-GLAD process. Detailed descriptions of these studies can be found in [9] and [31], respectively.

As expected from the difference in melting points between Ti and TiO_2, we do not observe significant differences from cross-sectional SEM images between the morphologies of both nano-sculpted films prepared for similar conditions (Figure 14a,b). Indeed, in both cases, T_s/T_m belongs to the Zone 1 regime (~0.19 vs. ~0.18, respectively). This suggests that our understanding of the growth of nano-sculpted thin films by RMS-GLAD based on the value of T_s/T_m is adapted independently of the chemistry of the studied compound. This observation is further supported by experimental data, revealing that self-diffusion from the bulk to the surface is initiated for temperatures higher than 400 K in TiO_2 [56]. This temperature is higher than the T_s value reached without intentional heating in our work (~373 K). We can, therefore, conclude that the growth mechanisms for Ti and TiO_2 are very close and that, as a consequence, the generated nano-sculpted films are similar. This is important as the simulation approach employed in our work is not yet adapted to the growth of oxide materials and, therefore, cannot be used to predict the morphology of such materials.

Nevertheless, this conclusion has to be moderated in certain situations such as that of pillar films shown in Figure 14c,d. In this case, because of the significantly lower deposition rate of the materials (2 vs. 10 nm/min for TiO_2 and Ti, respectively), the number of turns of the helical column increases consistently with the rotation speed, resulting in an elongated morphology [57]. Consequently, TiO_2 films have an elongated morphology by reducing the helical pitch (i.e., the height of one turn of a helix) due to the lower deposition rate, for similar rotation speed. However, the rotation speed limitation of the substrate holder (0.1°/s) does not allow the synthesis of films, presenting a morphology with large helicoidal pitches as that observed for Ti.

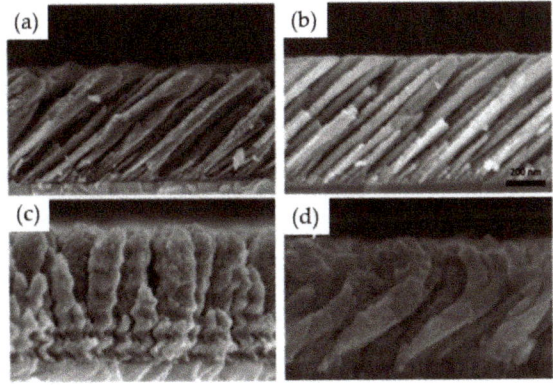

Figure 14. Cross-sectional SEM images of nano-columnar (**a**) TiO_2 and (**b**) Ti thin film deposited at 150 W, 0.13 Pa, and $\alpha = 85°$. Cross-sectional SEM images of helicoidal (**c**) TiO_2 and (**d**) Ti thin film deposited at 150 W, 0.13 Pa, $\alpha = 85°$, and $\phi_s = 0.1°/s$ (adapted from [50] and [9] with permission from Elsevier, 2017 and 2015, respectively).

As expected from the melting temperature difference between Mg and its corresponding oxide MgO, the story is different in this case. It indeed appears that the morphology of the sample is strongly affected by the addition of O_2 in the gas mixture (see Figure 15). The features of the columnar structures, i.e., the column width and the inter-columnar space, are reduced from (171 ± 18) to (56 ± 11) nm and from (120 ± 25) to (37 ± 16) nm, respectively, when increasing the flux of O_2 added to the discharge. Furthermore, the shape of the single columns is affected as well because pure Mg columns are strongly faceted, while they are no longer faceted when a small amount of O_2 is added to the discharge. However, β remains stable beyond the transition.

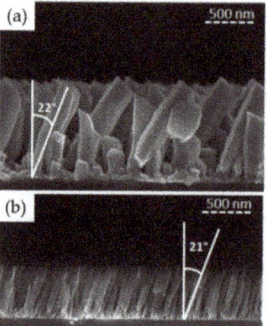

Figure 15. Cross-sectional SEM images of nano-columnar (**a**) Mg and (**b**) MgO thin film deposited at 50 W, 0.26 Pa, and α = 85°. The white angle accounts for the average columnar tilt angle (β) estimated by over 20 columns.

The modification of the chemistry of the deposited material actually impacts the diffusion kinetic of the growing material on its substrate. Indeed, in this case, T_m/T_s is significantly reduced for the MgO situation when compared to the Mg case (~0.10 vs. ~0.35, respectively). The deposition of MgO nano-sculpted films, thus, belongs to the Zone 1 regime of the SZM while the deposition of Mg belongs to the Zone 2 regime, as described above. This leads, for MgO, to a denser population of thinner grains acting as starting sites for the columns growth, thus explaining the evolution of the film microstructure in reactive conditions.

3.3. Growth Model for GLAD Deposited Thin Films

Based on the aforementioned studies, it is possible to adapt the empirical model for thin films deposited at normal incidence (α = 0°), proposed by Movchan and Demchishin, to deposition in an oblique angle configuration with different T_s (see Figure 16).

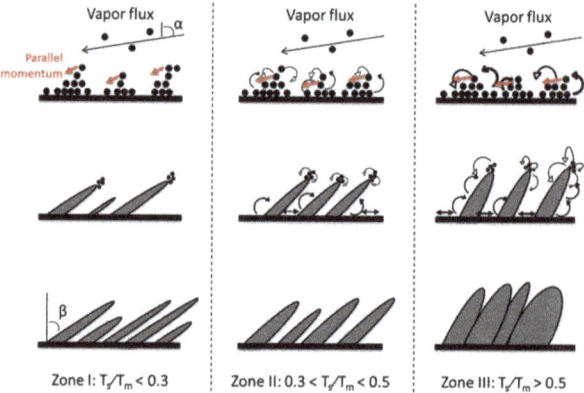

Figure 16. Growth models expected for oblique angle deposition at different substrate temperatures (reproduced from [50] with permission from Elsevier, 2017).

As already mentioned, the parallel momentum affects β, especially when the diffusion process is favored as it is the case under oblique incidence. In addition to that, the diffusion is also enhanced during deposition at high temperature. Thus, the homologous temperature limits of the adapted growth model for thin films under oblique incidence are the same as the normal incidence case.

In Zone 1 ($T_s/T_m < 0.3$), the diffusion process hardly takes place and the incident particles generally stay in the original deposition site, leading to a high number a thin columns producing small and numerous inter-columnar spaces.

For $0.3 < T_s/T_m < 0.5$, Zone 2 is reached, allowing hops up and hops down events after accommodation, until adatoms reach an "energetically favorable" site. The number of columns will, thus, decrease but become thicker because of the agglomeration process.

A further increase in T_s finally allows Zone 3 ($T_s/T_m > 0.5$) to be reached. The diffusion processes are increasingly enhanced and are now almost perpendicular to the substrate surface. Thus, the adatoms are able to reach the shadowed area, leading to the coalescence of the columns. The films porosity is thus highly reduced. More information about the time before accommodation can be found elsewhere [50].

3.4. Nano-Sculpted TiO₂ Films for Dye-Sensitized Solar Cell Applications

If the fundamental understanding of the observed phenomenon is the key element motivating our efforts, the applications of the developed materials in a meaningful field such as that of energy-related technologies are equally important. Therefore, the developed nano-sculpted thin films have been employed for energy-related applications such as for the design of novel hydrogen storage materials [32] or of novel photoanodes in DSSCs. The most important results obtained for the latter case will be summarized in the following pages.

3.4.1. Context

Considering current global concerns about the increase in energy consumption, the depletion of deeply used fossil fuels associated with the consequences of global warming makes the efficient use of renewable energies a major economic and environmental interest. Among all the studied renewable energy-based technologies, solar energy is definitely the most promising of them because of its abundance. In this context, dye-sensitized solar cells (DSSCs) are recognized as a potential low-cost photovoltaic solution owing to its many attractive advantages: Cheap production cost, tunable transparency, aesthetic capability, lightweight devices, but most importantly, good performances under low-illumination and high-temperature conditions [2,58,59].

The conventional architecture of a DSSC involves three key components: A porous array of n-type semiconductors (classically sintered TiO_2 nanoparticles), on which an electron-donor dye is adsorbed, both forming the photo-anode. The role of the dye is to absorb solar photons that causes its excitation, followed by electron injection into the conduction band of the semiconductor. Finally, the oxidized dye is regenerated by a redox liquid electrolyte, and the semiconductor transports the electron from the interface with the dye to the electrode [2,58,60]. The liquid electrolyte solution is then reduced at the counter electrode by electrons coming from the photo-anode via the external circuit. A scheme representing the different layers of DSSCs is shown in Figure 17.

In the first development of this technology in 1991, the efficiency was about 7–8% [2]. Most of the shortfall was due to the weak absorption of low energy photons by common dyes and to the high rate of electron–hole recombination at the semiconductor/dye and/or semiconductor/electrolyte interfaces. Hence, molecular engineering of the dyes and the development of new electrolyte mediators associated with the optimization of the device fabrication have enabled us to reach solar-to-electricity conversion efficiencies of 14% at the highest [61–64]. In addition, it is also accepted that the semiconductor composition and morphology strongly impact the performances of DSSCs, as recently reviewed by Maçaira et al. [65]. TiO_2 is considered nowadays as the best material because it is abundant, cheap, nontoxic, very stable under visible-light irradiation, and characterized by a wide bandgap, providing a high transmittance in the near UV-visible light region [66,67]. However, the conventionally used TiO_2-sintered nanoparticles electrodes are known to limit the charge transport by the structural disorder at the nanoparticles boundaries, enhancing electron scattering. This highly reduces the charge collection at the photo-anode, therefore limiting the current provided by the solar cell [9].

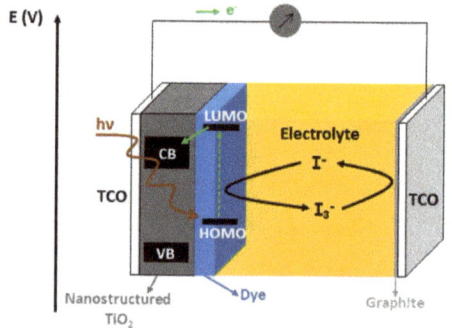

Figure 17. Schematic electron transport pathway in dye-sensitized solar cells (DSSCs).

Regarding the morphology, as well as the composition of the semi-conductor, being of major interest to allow a high dye uptake, as well as optimized charge transport efficiency, the recent advances performed by the use of nano-sculpted TiO$_2$-based photo-anodes is discussed in the following sections.

3.4.2. Nano-Sculpted TiO$_2$ Thin Films as Photoanode Material

In order to assess the potential of all morphologies that can be generated by RMS-GLAD, their respective aspect ratio, Γ (see Figure 11 for the definition), was evaluated from the analysis of SEM pictures. As expected, Γ depends on the film morphology: A dense film presents a 0 value while the largest value (0.3) is calculated for the slanted columnar morphology (SCM). According to the Γ parameter, the SCM morphology appears, therefore, to be the best candidate for optimizing the dye uptake capability [68]. This has been verified by assessing the desorption of dye to evaluate the effective specific surface area of the different films. Ru complex dye N719 was used as it allows a reversible grafting on the films. After dye absorption by dip coating of the thin films with fixed surface and thickness, the films were rinsed and immersed in a KOH solution in order to desorb the dye. The absorbance of the resulting solution was thus measured by UV-vis spectrophotometry to ultimately calculate the dye concentration using a calibration procedure. The SCM films allow the highest absorption in agreement with the expectations from the Γ values. The highest specific surface area of 86 m^2/g was obtained for films deposited with $\alpha = 85°$, while the optimized film thickness was evaluated at 3.5 µm with a surface area enhancement around 200 m$^2_{N719}$/m$^2_{substrate}$ [68].

The crystalline structure of the deposited films was investigated by X-ray diffraction, which reveals an anatase phase for all generated morphologies. To obtain space-resolved information, electron diffraction during TEM experiments was performed, showing that the films are not homogeneously crystallized: An amorphous phase is deduced at the substrate interface while the anatase phase was detected near the surface region. This late crystallization at the end of the film growth can be explained by a gradual heating of the substrate by ions bombardment, surface reactions, and intense IR radiation emanating from the target [27].

In view of the utilization of these films in DSSCs, a homogeneous crystallization of the material is necessary. The usual strategies allowing us to improve the crystalline quality of RMS-deposited films have, therefore, been evaluated, namely biasing the substrate, the substrate heating during the deposition, and the post-annealing of the synthesized films. The data, fully described in [9], revealed that a substrate polarization allows the control of the crystalline phase from pure anatase to pure rutile, but leads to the densification of the thin films while both heating procedures allow a better crystallization into the anatase phase without significantly affecting the film morphology (see Figure 18).

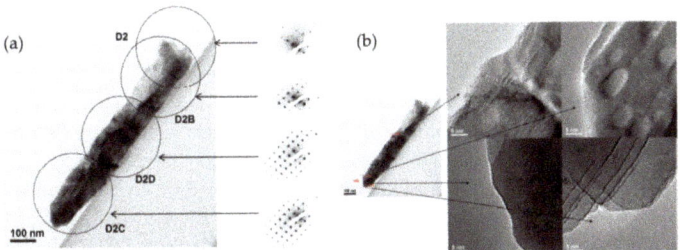

Figure 18. High-resolution TEM picture of an isolated column from a slanted columnar TiO$_2$ thin film annealed 2 h at 773 K; (**a**) the corresponding electron diffraction patterns; (**b**) high-magnification pictures at various locations.

These films have then been utilized as the photo-anode in DSSCs. SCMs synthesized for different α and thicknesses were investigated. A reference photo-anode based on conventional TiO$_2$ nanoparticles powder was systematically built to validate the utilization of our thin films. The photovoltaic parameters, i.e., the fill factor (FF), open-circuit voltage (V_{oc}), short-circuit current (J_{sc}), and efficiency (η), are presented in Figure 19.

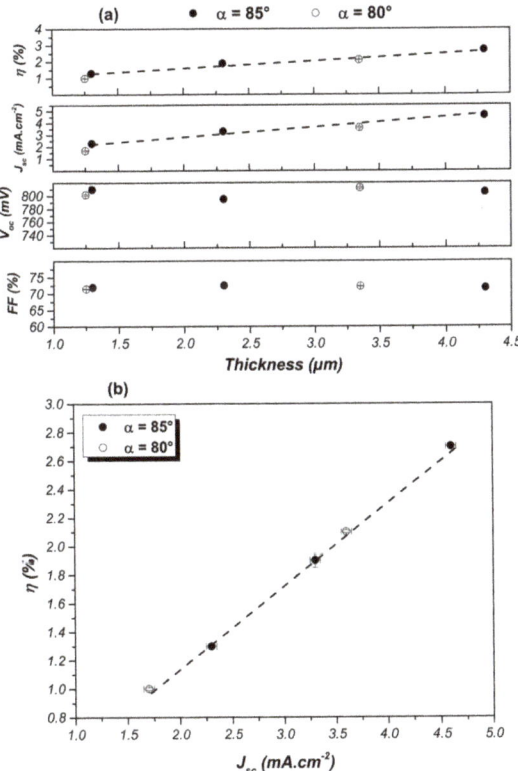

Figure 19. (**a**) Photovoltaic performances of liquid DSSCs integrating a slanted columns-based TiO$_2$ film as the photo-anode and according to the thickness of the latter; (**b**) plot of the cell efficiency according to the corresponding J_{sc}.

First, FF and V_{oc} parameters are constant around 72% and ~800 mV, respectively. These observations are not surprising, as the combination of MS and GLAD allows well-adherent and -ordered sculpted TiO_2 thin films to be generated onto the FTO layer, ensuring a good contact. All photo-anodes were post-annealed under an identical procedure, leading to the same crystalline structure.

Furthermore, Figure 19b reveals that the cell efficiency linearly evolves with the current density produced by the cell whatever the angle of deposition. This means that the quantity of adsorbed dye through the corresponding J_{sc} is mainly responsible for the overall efficiency of our DSSCs. However, the reference cell based on TiO_2 nanoparticles is characterized by the relatively same FF and V_{oc} (69% and 752 mV, respectively), while the J_{sc} and η are around 20 mA·cm^{-2} and 10.7%, respectively, meaning that J_{sc} is the parameter responsible for the four-times-higher efficiency. Consequently, the slanted columnar morphology allows good charge transport while the density of adsorbed dye remains the critical parameter [35].

3.4.3. A Nano-Sculpted TiO_2/TiO_2 Nanoparticles Hybrid Approach

The easiest way to increase the dye absorption density would be to use thicker photo-anodes. Nevertheless, this option is limited by the RMS process, which allows for the synthesis of less than ~1 µm films typically. On the other hand, the low deposition rate of the nano-sculpted films (~2 nm/min typically) would also be a limiting factor. Therefore, as an alternative, it has been decided to impregnate the nano-sculpted TiO_2 films with a solution of anatase TiO_2 nanoparticles (Solaronix®, Aubonne, Switzerland) ~20 nm by spin coating. In this way, a hierarchical structure is formed and the voids between the columns are exploited to ultimately increase the dye molecule absorption by the nanoparticles, allowing for a better dye uptake. A cross-sectional SEM image of this hybrid system is shown in Figure 20.

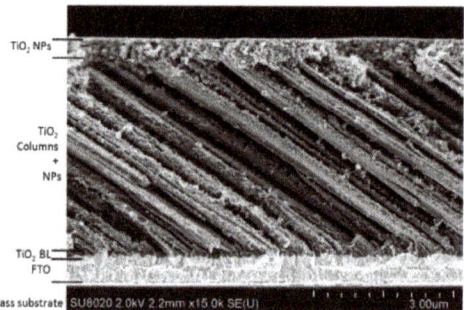

Figure 20. Cross-sectional SEM picture of a slanted columnar thin film (4.3 µm) after spin coating by a TiO_2 nanoparticles solution.

The photovoltaic performances of the DSSCs based on this hybrid photo-anode have been evaluated, and the results are presented in Figure 21a. First, it appears that the incorporation of NPs significantly increases the absorbance of the film to a value of 0.0518 in comparison with the value of 0.0243 for the SCM thin films (4.3 µm) alone. The incorporation of NPs significantly improves the J_{sc} value from 4.6 mA/cm^2 for a "simple" columnar thin film (4.3 µm) to 10.6 mA/cm^2 when adding the NPs (4.3 µm). The differences in terms of light absorption appear to be the main explanation for the different J_{sc} values that are measured, whereas for the screen-printed NPs film (9.6 µm), which absorbs the light one order of magnitude better than the SCM photo-anode, the J_{sc} value and η are only improved by a factor of 4. These results suggest that for the screen-printed NPs, in spite of the amount of dye adsorbed, which is much larger, many of the generated charge carriers are not collected, which is in line with the structural disorder of NPs that contributes to the loss of charges.

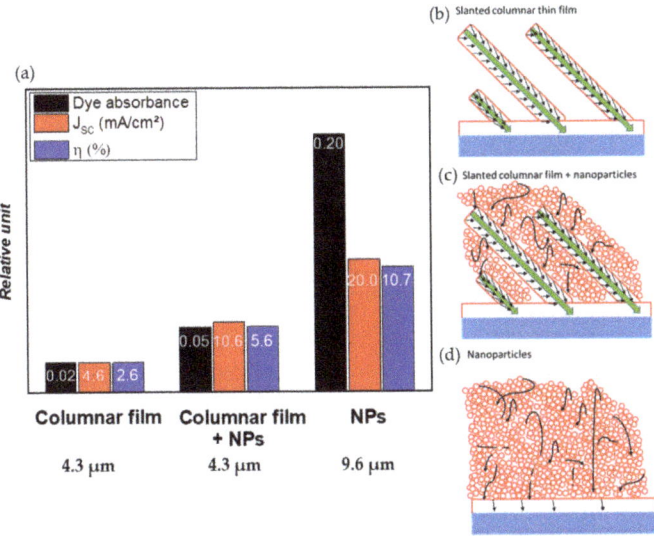

Figure 21. (a) Photovoltaic performances of DSSCs based on various photo-anode architectures. Schematic representation of the electron transfer occurring in photo-anode based on: (b) Slanted columnar thin film; (c) the combination of slanted columns and nanoparticles; (d) nanoparticulate thin films.

These results can be explained by the synergistic effect between the monocrystalline column, which act as extremely good conductive media for the electron, while the additional nanoparticles in the inter-columnar voids help to graft more dye and, therefore, to increase the amount of generated photoelectrons. On the contrary, for a conventional nanoparticle anode, the adsorption of the dye is of very good quality, but the generated electrons are lost due to the already mentioned defective structure of the anode. These mechanisms are depicted in Figure 21b–d. It is worth stressing that the comparison between the different photoanode architectures was not made with films of the same thickness, owing to the following assumption: It is accepted that J_{sc} linearly increases as a function of the thickness of the nanoparticle-based thin film, before reaching a plateau and finally decreasing [69]. As the thicknesses considered in our work (<10 μm) still belong to the linear region, the recombination rate is not yet a limiting parameter, allowing our discussions and conclusions to be meaningful.

We can, therefore, conclude that, in this case of study, the integration of NPs allows an increase in the light absorption by improving the dye impregnation, while the nano-sculpted thin film allows an efficient collection and transfer of charges, avoiding the recombination reactions. Generating this synergistic effect between the nanoparticles and the single crystalline columns seems to be a good strategy to ultimately increase the overall efficiency of DSSCs.

4. Conclusions

This work summarizes our recent research related to the development of nano-sculpted thin films by magnetron-sputtering-related technologies and to their use in energy-related applications. We first describe our understanding of the growth mechanism associated with the novel utilization of the glancing-angle geometry in magnetron sputtering processes, the MS-GLAD process. The synthesis of model nano-sculpted Ti and Mg films was investigated by using a joint experimental–modeling approach based on kMC simulations implemented in the NASCAM code.

Based on the different morphological properties of Ti and Mg coatings grown for similar experimental conditions, it appears that the homologous temperature as defined in structural zone models, T_s/T_m, is one of the key parameters to finely control the growth of nano-sculpted coatings by

MS-GLAD. When comparing the two considered metals, this parameter is different enough to allow for different growth regimes, from Zone 1 for Ti to Zone 2 for Mg. Basically, this parameter mainly defines the importance of the adatom diffusion processes that is, at room temperature, negligible in the Ti case (the ballistic deposition approximation is sufficient), while events such as hops up and hops down from one atomic layer to another should be taken into account to accurately describe the growth of Mg nano-sculpted thin films. This rationalization based on the homologous temperature of the deposited material can even been extended when considering a different chemistry of the system, i.e., an oxidation of the deposited material. Indeed, in such a situation, it is shown that, as expected by the values of the homologous temperature, Ti and TiO_2 behave almost similarly while a strong impact is observed for Mg.

The other key parameters are related to the collimated nature of the depositing flux and on its impact on the shadowing effect, which is the basic effect when GLAD geometry is considered. Therefore, it has been demonstrated that the deposition pressure, which strongly affects the collimated character of the depositing flux through the mean free path of the particles, has to be low enough (<0.26 Pa) to trigger the formation of the different nano-sculpted structures.

In comparison with the conventional combination of GLAD with evaporation, it is shown that the utilization of magnetron sputtering in given conditions allows for a good crystallization of the deposition material, which is important in many applications. In particular, for TiO_2 thin films, we demonstrate that anatase monocrystalline-like nanocolumns-based thin films can be synthesized.

From an application point of view, nano-sculpted TiO_2 coatings were integrated into the photo-anode of dye-sensitized solar cells (DSSCs). First, it appears that the devices based on nano-sculpted thin films outperform nanoparticles-based DSSCs both in terms of charge harvesting and charge recombination. However, the photo-anode thickness drastically affects the cell performances, indicating that the critical parameter is the adsorbed dye density. This problem has been addressed by combing the nano-sculpted TiO_2 films with a spin-coated TiO_2 nanoparticles solution. This hybrid system demonstrates a synergetic effect between the columnar thin film and the absorbed nanoparticles, which significantly improved the efficiency of the DSSCs by simultaneously enhancing the charge transport and the quantity of adsorbed dye molecules.

We believe that the development of novel magnetron sputtering approaches to design materials presenting a well-defined morphology at the nanoscale consists of an important opportunity for this well-established technology.

Supplementary Materials: The following are available online at http://www.mdpi.com/2079-4991/10/10/2039/s1, Figures S1 and S2: Cross-sectional SEM images with the corresponding simulations of Ti and Mg thin films synthesized at various angles of deposition; Figure S3: Angular distribution of particles calculated by SIMTRA for various angles of deposition; Figure S4: Cross-sectional SEM images with the corresponding simulations of Mg thin films synthesized at various deposition pressures.

Author Contributions: Conceptualization, R.S.; methodology, A.P. and R.S.; software, A.P.; validation, A.P. and R.S.; formal analysis, A.P.; investigation, A.P.; resources, A.P.; data curation, A.P. and R.S.; writing—original draft preparation, A.P.; writing—review and editing, R.S.; visualization, A.P.; supervision, R.S.; project administration, R.S.; funding acquisition, A.P. and R.S. All authors have read and agreed to the published version of the manuscript.

Funding: A.P. is supported by a F.R.I.A grant of National Fund for Scientific Research (FNRS – Belgium).

Acknowledgments: The authors thank J. Dervaux, P.-A. Cormier, H. Liang, X. Geng, N. Szuwarski, Y. Pellegrin, E. Gautron, M. Boujtita, S. Konstantinidis and F. Odobel for their participation in the various studies discussed in this paper.

Conflicts of Interest: The authors declare no conflict of interest.

References

1. Hawkeye, M.M.; Taschuk, M.T.; Brett, M.J. *Glancing Angle Deposition of Thin Films*; Wiley: Hoboken, NJ, USA, 2014.
2. O'Regan, B.; Grätzel, M. A low-cost, high-efficiency solar cell based on dye-sensitized colloidal TiO_2 films. *Nature* **1991**, *353*, 737–740. [CrossRef]

3. Forro, L.; Chauvet, O.; Emin, D.; Zuppiroli, L. High mobility n-type charge carriers in large single crystals of anatase (TiO$_2$). *J. Appl. Phys.* **1994**, *75*, 633–635. [CrossRef]
4. Agrell, H.G.; Boschloo, G.; Hagfeldt, A. Conductivity Studies of Nanostructured TiO$_2$ Films Permeated with Electrolyte. *J. Phys. Chem. B* **2004**, *108*, 12388–12396. [CrossRef]
5. Mor, G.K.; Shankar, K.; Paulose, M.; Varghese, O.K.; Grimes, C.A. Use of Highly-Ordered TiO2 Nanotube Arrays in Dye-Sensitized Solar Cells. *Nano Lett.* **2006**, *6*, 215–218. [CrossRef]
6. Du, S.; Koenigsmann, C.; Sun, S. One-dimensional nanostructures for PEM fuel cell applications. In *Hydrogen and Fuel Cells Primers*; Pollet, B., Ed.; Elsevier: Amsterdam, The Netherlands, 2017.
7. Xie, Z.; Henry, B.; Kirov, K.; Smith, H.; Barkhouse, A.; Grovenor, C.; Assender, H.; Briggs, G.; Webster, G.; Burn, P.L.; et al. Study of the effect of changing the microstructure of titania layers on composite solar cell performance. *Thin Solid Films* **2006**, *511*, 523–528. [CrossRef]
8. Govardhan Reddy, K.; Deepak, T.G.; Anjusree, G.S.; Thomas, S.; Vadukumpully, S.; Subramanian, K.R.V.; Nair, S.V.; Nair, A.S. On Global Energy Scenario, Dye-sensitized Solar Cells and the Promise of Nanotechnology Optoelectron. *Adv. Mater. Rapid Commun.* **2010**, *4*, 1166–1169.
9. Dervaux, J.; Cormier, P.-A.; Konstantinidis, S.; Di Ciuccio, R.; Coulembier, O.; Dubois, P.; Snyders, R. Deposition of porous titanium oxide thin films as anode material for dye sensitized solar cells. *Vacuum* **2015**, *114*, 213–220. [CrossRef]
10. Wu, W.-Q.; Feng, H.-L.; Rao, H.-S.; Xu, Y.-F.; Kuang, D.-B.; Su, C.-Y. Maximizing omnidirectional light harvesting in metal oxide hyperbranched array architectures. *Nat. Commun.* **2014**, *5*, 3968. [CrossRef]
11. Krumpmann, A. Anodized TiO$_2$ nanotubes as a photoelectrode material for solid-state dye-sensitized solar cells. PhD'Thesis, University of Mons, Mons, Belgium, 2018.
12. Colgan, M.; Djurfors, B.; Ivey, D.; Brett, M. Effects of annealing on titanium dioxide structured films. *Thin Solid Films* **2004**, *466*, 92–966. [CrossRef]
13. Sander, M.S.; Côté, M.J.; Gu, W.; Kile, B.M.; Tripp, C.P. Template-Assisted Fabrication of Dense, Aligned Arrays of Titania Nanotubes with Well-Controlled Dimensions on Substrates. *Adv. Mater.* **2004**, *16*, 2052–2057. [CrossRef]
14. Shin, H.; Jeong, D.-K.; Lee, J.; Sung, M.M.; Kim, J. Formation of TiO$_2$ and ZrO$_2$ Nanotubes Using Atomic Layer Deposition with Ultraprecise Control of the Wall Thickness. *Adv. Mater.* **2004**, *16*, 1197–1200. [CrossRef]
15. Law, M.; Greene, L.E.; Johnson, J.C.; Saykally, R.; Yang, P. Nanowire dye-sensitized solar cells. *Nat. Mater.* **2005**, *4*, 455–459. [CrossRef] [PubMed]
16. Sadeghzadeh-Attar, A.; Ghamsari, M.S.; Hajiesmaeilbaigi, F.; Mirdamadi, S.; Katagiri, K.; Koumoto, K. Sol–gel template synthesis and characterization of aligned anatase-TiO$_2$ nanorod arrays with different diameter. *Mater. Chem. Phys.* **2009**, *113*, 856–860. [CrossRef]
17. Feng, X.; Zhu, K.; Frank, A.J.; Grimes, C.A.; Mallouk, T.E. Rapid Charge Transport in Dye-Sensitized Solar Cells Made from Vertically Aligned Single-Crystal Rutile TiO$_2$ Nanowires. *Angew. Chem. Int. Ed.* **2012**, *51*, 2727–2730. [CrossRef] [PubMed]
18. Liu, N.; Chen, X.; Zhang, J.; Schwank, J.W. A review on TiO$_2$-based nanotubes synthesized via hydrothermal method: Formation mechanism, structure modification, and photocatalytic applications. *Catal. Today* **2014**, *225*, 34–51. [CrossRef]
19. Grill, A. *Cold Plasma Materials Fabrication*; Institute of Electrical and Electronics Engineers (IEEE), Wiley: Piscataway, NJ, USA, 1994.
20. Ershov, A.; Pekker, L. Model of d.c. magnetron reactive sputtering in Ar-O$_2$ gas mixtures. *Thin Solid Films* **1996**, *289*, 140–146. [CrossRef]
21. Safi, I. Recent aspects concerning DC reactive magnetron sputtering of thin films: A review. *Surf. Coatings Technol.* **2000**, *127*, 203–218. [CrossRef]
22. Cormier, P.-A.; Balhamri, A.; Thomann, A.-L.; Dussart, R.; Semmar, N.; Lecas, T.; Snyders, R.; Konstantinidis, S. Titanium oxide thin film growth by magnetron sputtering: Total energy flux and its relationship with the phase constitution. *Surf. Coat. Technol.* **2014**, *254*, 291–297. [CrossRef]
23. Bräuer, G.; Szyszka, B.; Vergöhl, M.; Bandorf, R. Magnetron sputtering – Milestones of 30 years. *Vacuum* **2010**, *84*, 1354–1359. [CrossRef]
24. Young, N.O.; Kowal, J. Optically Active Fluorite Films. *Nature*, 1959; *183*, 104–105.
25. Michalcik, Z.; Horakova, M.; Spatenka, P.; Klementová, Š.; Zlámal, M.; Martin, N. Photocatalytic Activity of Nanostructured Titanium Dioxide Thin Films. *Int. J. Photoenergy* **2012**, *2012*, 1–8. [CrossRef]

26. Robbie, K.; Brett, M.J. Sculptured thin films and glancing angle deposition: Growth mechanics and applications. *J. Vac. Sci. Technol. A* **1997**, *15*, 1460–1465. [CrossRef]
27. Cormier, P.-A.; Thomann, A.-L.; Dolique, V.; Balhamri, A.; Dussart, R.; Semmar, N.; Lecas, T.; Brault, P.; Snyders, R.; Konstantinidis, S. IR emission from the target during plasma magnetron sputter deposition. *Thin Solid Films* **2013**, *545*, 44–49. [CrossRef]
28. Sit, J.C.; Vick, D.; Robbie, K.; Brett, M.J. Thin Film Microstructure Control Using Glancing Angle Deposition by Sputtering. *J. Mater. Res.* **1999**, *14*, 1197–1199. [CrossRef]
29. García-Martín, J.M.; Alvarez, R.; Romero-Gomez, P.; Cebollada, A.; Palmero, A. Tilt angle control of nanocolumns grown by glancing angle sputtering at variable argon pressures. *Appl. Phys. Lett.* **2010**, *97*, 173103. [CrossRef]
30. Anders, A. A structure zone diagram including plasma-based deposition and ion etching. *Thin Solid Films* **2010**, *518*, 4087–4090. [CrossRef]
31. Geng, X.; Liang, H.; Li, W.; Panepinto, A.; Thiry, D.; Chen, M.; Snyders, R. Experimental evaluation of the role of oxygen on the growth of MgOx nano-sculpted thin films synthesized by reactive magnetron sputtering combined with glancing angle deposition. *Thin Solid Films*. submitted.
32. Liang, H.; Geng, X.; Li, W.; Panepinto, A.; Thiry, D.; Chen, M.; Snyders, R. Experimental and Modeling Study of the Fabrication of Mg Nano-Sculpted Films by Magnetron Sputtering Combined with Glancing Angle Deposition. *Coatings* **2019**, *9*, 361. [CrossRef]
33. Ganciu, M.; Konstantinidis, S.; Paint, Y.; Dauchot, J.P.; Hecq, M.; De Poucques, L.; Vašina, P.; Meško, M.; Imbert, J.C.; Bretagne, J.; et al. Preionised pulsed magnetron discharges for ionised physical vapour deposition. *J. Optoelectron. Adv. Mater.* **2005**, *7*, 2481–2484.
34. Panepinto, A.; Michiels, M.; Dürrschnabel, M.T.; Molina-Luna, L.; Bittencourt, C.; Cormier, P.A.; Snyders, R. Synthesis of Anatase (Core)/Rutile (Shell) Nanostructured TiO_2 Thin Films by Magnetron Sputtering Methods for Dye-Sensitized Solar Cell Applications. *ACS Appl. Energy Mater.* **2020**, *3*, 759–767. [CrossRef]
35. Cormier, P.-A.; Dervaux, J.; Szuwarski, N.; Pellegrin, Y.; Odobel, F.; Gautron, E.; Boujtita, M.; Snyders, R.; Boujita, M. Single Crystalline-like and Nanostructured TiO2 Photoanodes for Dye Sensitized Solar Cells Synthesized by Reactive Magnetron Sputtering at Glancing Angle. *J. Phys. Chem. C* **2018**, *122*, 20661–20668. [CrossRef]
36. Bortz, A.; Kalos, M.; Lebowitz, J. A new algorithm for Monte Carlo simulation of Ising spin systems. *J. Comput. Phys.* **1975**, *17*, 10–18. [CrossRef]
37. Claassens, C.H.; Hoffman, M.J.H.; Terblans, J.; Swart, H.C. Kinetic Monte Carlo Simulation of the Growth of Various Nanostructures through Atomic and Cluster Deposition: Application to Gold Nanostructure Growth on Graphite. *J. Phys. Conf. Ser.* **2006**, *29*, 185–189. [CrossRef]
38. Meakin, P.; Krug, J. Three-dimensional ballistic deposition at oblique incidence. *Phys. Rev. A* **1992**, *46*, 3390–3399. [CrossRef] [PubMed]
39. Smy, T.; Vick, D.; Brett, M.J.; Dew, S.K.; Wu, A.T.; Sit, J.C.; Harris, K.D. Three-dimensional simulation of film microstructure produced by glancing angle deposition. *J. Vac. Sci. Technol. A* **2000**, *18*, 2507. [CrossRef]
40. Lucas, S.; Moskovkin, P. Simulation at high temperature of atomic deposition, islands coalescence, Ostwald and inverse Ostwald ripening with a general simple kinetic Monte Carlo code. *Thin Solid Films* **2010**, *518*, 5355–5361. [CrossRef]
41. NASCAM (NAnoSCAle Modeling). Available online: https://www.unamur.be/sciences/physique/ur/larn/logiciels/nascam (accessed on 5 May 2020).
42. Ziegler, J.F.; Ziegler, M.D.; Biersack, J.P. SRIM—The stopping and range of ions in matter. *Nucl. Instrum. Methods Phys. Res. B* **2010**, *268*, 1818–1823. [CrossRef]
43. Godinho, V.; Moskovkin, P.; Álvarez, R.; Caballero-Hernández, J.; Schierholz, R.; Bera, B.; Demarche, J.; Palmero, A.; Fernández, A.; Lucas, S. On the formation of the porous structure in nanostructured a-Si coatings deposited by dc magnetron sputtering at oblique angles. *Nanotechnology* **2014**, *25*, 355705. [CrossRef]
44. Thornton, J.A. Influence of apparatus geometry and deposition conditions on the structure and topography of thick sputtered coatings. *J. Vac. Sci. Technol.* **1974**, *11*, 666–670. [CrossRef]
45. Hussla, I.; Enke, K.; Grunwald, H.; Lorenz, G.; Stoll, H. In situ silicon-wafer temperature measurements during RF argon-ion plasma etching via fluoroptic thermometry. *J. Phys. D Appl. Phys.* **1987**, *20*, 889–896. [CrossRef]
46. Kersten, H.; Deutsch, H.; Steffen, H.; Kroesen, G.; Hippler, R. The energy balance at substrate surfaces during plasma processing. *Vacuum* **2001**, *63*, 385–431. [CrossRef]
47. Kersten, H.; Rohde, D.; Steffen, H.; Deutsch, H.; Hippler, R.; Swinkels, G.; Kroesen, G. On the determination of energy fluxes at plasma–surface processes. *Appl. Phys. A* **2001**, *72*, 531–540. [CrossRef]

48. Movchan, B.A.; Demchishin, A.V. Structure and Properties of Thick Condensates of Nickel, Titanium, Tungsten, Aluminium Oxides, and Zirconium Dioxide in Vacuum. *Phys. Metal. Metallog.* **2014**, *28*, 653–663.
49. Jain, I.; Lal, C.; Jain, A. Hydrogen storage in Mg: A most promising material. *Int. J. Hydrogen Energy* **2010**, *35*, 5133–5144. [CrossRef]
50. Dervaux, J.; Cormier, P.-A.; Moskovkin, P.; Douheret, O.; Konstantinidis, S.; Lazzaroni, R.; Lucas, S.; Snyders, R. Synthesis of nanostructured Ti thin films by combining glancing angle deposition and magnetron sputtering: A joint experimental and modeling study. *Thin Solid Films* **2017**, *636*, 644–657. [CrossRef]
51. Abelmann, L.; Lodder, C. Oblique evaporation and surface diffusion. *Thin Solid Films* **1997**, *305*, 1–21. [CrossRef]
52. Rohlf, J.W. Modern physics from [alpha] to Z0. Wiley: Hoboken, NJ, USA, 1994.
53. Blocher, J.M.; Campbell, I.E. Vapor Pressure of Titanium. *J. Am. Chem. Soc.* **1949**, *71*, 4040–4042. [CrossRef]
54. Hawkeye, M.M.; Brett, M.J. Glancing angle deposition: Fabrication, properties, and applications of micro- and nanostructured thin films. *J. Vac. Sci. Technol. A* **2007**, *25*, 1317. [CrossRef]
55. Dick, B.; Brett, M.J.; Smy, T. Controlled growth of periodic pillars by glancing angle deposition. *J. Vac. Sci. Technol. B Microelectron. Nanometer Struct.* **2003**, *21*, 23. [CrossRef]
56. Henderson, M.A. A surface perspective on self-diffusion in rutile TiO_2. *Surf. Sci.* **1999**, *419*, 174–187. [CrossRef]
57. Yao, K.-S.; Chen, Y.-C.; Chao, C.-H.; Wang, W.-F.; Lien, S.-Y.; Shih, H.C.; Chen, T.-L.; Weng, K.-W. Electrical enhancement of DMFC by Pt–M/C catalyst-assisted PVD. *Thin Solid Films* **2010**, *518*, 7225–7228. [CrossRef]
58. Hagfeldt, A.; Boschloo, G.; Sun, L.; Kloo, L.; Pettersson, H. Dye-Sensitized Solar Cells. *Chem. Rev.* **2010**, *110*, 6595–6663. [CrossRef]
59. Grätzel, M. Conversion of sunlight to electric power by nanocrystalline dye-sensitized solar cells. *J. Photochem. Photobiol. A Chem.* **2004**, *164*, 3–14.
60. Freitag, M.; Teuscher, J.; Saygili, Y.; Zhang, X.; Giordano, F.; Liska, P.; Hua, J.; Zakeeruddin, S.M.; Moser, J.-E.; Grätzel, M.; et al. Dye-sensitized solar cells for efficient power generation under ambient lighting. *Nat. Photonics* **2017**, *11*, 372–378. [CrossRef]
61. Zhang, L.; Yang, X.; Wang, W.; Gurzadyan, G.G.; Li, J.; Li, X.; An, J.; Yu, Z.; Wang, H.; Cai, B.; et al. 13.6% Efficient Organic Dye-Sensitized Solar Cells by Minimizing Energy Losses of the Excited State. *ACS Energy Lett.* **2019**, *4*, 943–951. [CrossRef]
62. Kakiage, K.; Aoyama, Y.; Yano, T.; Oya, K.; Fujisawa, J.-I.; Hanaya, M. Highly-efficient dye-sensitized solar cells with collaborative sensitization by silyl-anchor and carboxy-anchor dyes. *Chem. Commun.* **2015**, *51*, 15894–15897. [CrossRef]
63. Yella, A.; Lee, H.-W.; Tsao, H.N.; Yi, C.; Chandiran, A.K.; Nazeeruddin, M.K.; Diau, E.W.-G.; Yeh, C.-Y.; Zakeeruddin, S.M.; Grätzel, M. Porphyrin-Sensitized Solar Cells with Cobalt (II/III)-Based Redox Electrolyte Exceed 12 Percent Efficiency. *Science* **2011**, *334*, 629–634. [CrossRef]
64. Mathew, S.; Yella, A.; Gao, P.; Humphry-Baker, R.; Curchod, B.F.E.; Astani, N.A.; Tavernelli, I.; Rothlisberger, U.; Nazeeruddin, K.; Grätzel, M. Dye-sensitized solar cells with 13% efficiency achieved through the molecular engineering of porphyrin sensitizers. *Nat. Chem.* **2014**, *6*, 242–247. [CrossRef] [PubMed]
65. Maçaira, J.; Andrade, L.; Mendes, A. Review on nanostructured photoelectrodes for next generation dye-sensitized solar cells. *Renew. Sustain. Energy Rev.* **2013**, *27*, 334–349. [CrossRef]
66. Mardare, D.; Tasca, M.; Delibas, M.; Rusu, G. On the structural properties and optical transmittance of TiO_2 r.f. sputtered thin films. *Appl. Surf. Sci.* **2000**, *156*, 200–206. [CrossRef]
67. Diebold, U. The surface science of titanium dioxide. *Surf. Sci. Rep.* **2003**, *48*, 53–229. [CrossRef]
68. Dervaux, J. Synthesis of nanostructured TiO_2 thin films by reactive magnetron sputtering in glancing angle configuration for dye-sensitized solar cell applications. PhD Thesis, University of Mons, Mons, Belgium, 2017.
69. Kang, M.G.; Ryu, K.S.; Chang, S.H.; Park, N.G.; Hong, J.S.; Kim, K.J. Dependence of TiO_2 Film Thickness on Photocurrent-Voltage Characteristics of Dye-Sensitized Solar Cells. *Bull. Korean Chem. Soc.* **2004**, *25*, 742–744.

Publisher's Note: MDPI stays neutral with regard to jurisdictional claims in published maps and institutional affiliations.

© 2020 by the authors. Licensee MDPI, Basel, Switzerland. This article is an open access article distributed under the terms and conditions of the Creative Commons Attribution (CC BY) license (http://creativecommons.org/licenses/by/4.0/).

Article

Significant Performance and Stability Improvements of Low-Temperature IGZO TFTs by the Formation of In-F Nanoparticles on an SiO$_2$ Buffer Layer

Ho-young Jeong [1,2], Seung-hee Nam [2], Kwon-shik Park [2], Soo-young Yoon [2], Chanju Park [1] and Jin Jang [1,*]

1. Advanced Display Research Center, Department of Information Display, Kyung Hee University, Dongdaemun-gu, Seoul 130-701, Korea; hyjeong@lgdisplay.com (H.-y.J.); cjpark@tft.khu.ac.kr (C.P.)
2. LG Display R&D Center, 245, Lg-ro, Paju-si, Gyeonggi-do 413-811, Korea; shnam@lgdisplay.com (S.-h.N.); pius@lgdisplay.com (K.-s.P.); syyoon@lgdisplay.com (S.-y.Y.)
* Correspondence: jjang@khu.ac.kr

Received: 12 May 2020; Accepted: 9 June 2020; Published: 15 June 2020

Abstract: We report the performance improvement of low-temperature coplanar indium–gallium–zinc–oxide (IGZO) thin-film transistors (TFTs) with a maximum process temperature of 230 °C. We treated F plasma on the surface of an SiO$_2$ buffer layer before depositing the IGZO semiconductor by reactive sputtering. The field-effect mobility increases from 3.8 to 9.0 cm^2 V^{-1}·s^{-1}, and the threshold voltage shift (ΔV_{th}) under positive-bias temperature stress decreases from 3.2 to 0.2 V by F-plasma exposure. High-resolution transmission electron microscopy and atom probe tomography analysis reveal that indium fluoride (In-F) nanoparticles are formed at the IGZO/buffer layer interface. This increases the density of the IGZO and improves the TFT performance as well as its bias stability. The results can be applied to the manufacturing of low-temperature coplanar oxide TFTs for oxide electronics, including information displays.

Keywords: low-temperature coplanar IGZO TFT; bias stability; In-F nanoparticles

---

1. Introduction

Indium–gallium–zinc–oxide (IGZO) thin-film transistors (TFTs) are widely used as a TFT backplane for active-matrix liquid-crystal displays (AMLCDs) and active-matrix organic light-emitting diode (AMOLED) displays due to their higher mobility and smaller subthreshold swing compared to amorphous silicon (a-Si) TFTs [1–4]. To realize the future of displays with flexible and transparent functions, transparent polyimide (PI) and poly(ethersulfone) (PES) could be used, and a low-temperature process (~230 °C) is required. The advantage of oxide TFTs is that they require a low-temperature process, but a relatively high-temperature process at about 350 °C is being used for the mass production of displays. This is mainly due to the improvement of the mobility and bias stability of oxide TFTs [5,6]. It is noted that the quality of SiO$_2$ depends on its substrate process temperature; the density and leakage through SiO$_2$ and electrical breakdown voltage could be improved by increasing the deposition temperature of SiO$_2$. The buffer SiO$_2$ of coplanar structures and the gate dielectric SiO$_2$ of back-channel-etch (BCE) structures are usually deposited at 300 to 400 °C by plasma-enhanced chemical vapor deposition (PECVD) for display applications [7,8]. The coplanar structure of IGZO TFTs has the advantage of negligible overlap capacitance between the gate and source/drain electrodes, and thus, the RC (resistance–capacitance product) delay can be remarkably reduced for display applications [5–7] [9–11]. In addition, the coplanar oxide TFT is a strong candidate for low-temperature devices because it could not use high-temperature SiO$_2$ as a gate insulator (GI) because of the damage this would cause to IGZO during GI deposition [12]. In order words, coplanar oxide TFTs with only a high-temperature buffer layer have good performance to be used in AMOLED televisions [13,14], and the coplanar oxide TFTs with a low-temperature buffer layer could be widely used as low-temperature TFTs, if the TFTs improve their performance. Note that low-temperature coplanar IGZO TFTs suffer from lower mobility and higher threshold voltage shift (ΔV_{th}) under positive-bias temperature stress (PBTS), and improving PBTS stability is important for the development of stable IGZO TFTs for AMOLED displays, given that driving TFTs in the OLED pixel are positively biased, and even a small change in the V_{th} can deteriorate image quality [15,16].

In this work, we studied the effect of F-plasma treatment on low-temperature SiO_2 to improve the performance and bias stability, PBTS, of the low-temperature coplanar IGZO TFT. The physical damage by ion bombardment on the SiO_2 buffer layer during the sputter deposition of IGZO might be one of the main reasons for the generation of trap sites in low-temperature SiO_2 [17]. To reduce the trap sites generated during the sputtering process, we performed F-plasma treatment on the SiO_2 buffer because F is known to passivate traps in SiO_2 as well as in IGZO [18–22]. Methods such as dynamic secondary-ion mass spectroscopy (D-SIMS), high-resolution transmission electron microscopy (HR-TEM) and APT (atom probe tomography) were used to investigate the relationships between the process temperature of the buffer layer, the performance of the TFTs and the effect of plasma treatment on the buffer layer.

2. Materials and Methods

The cross-sectional view of the coplanar IGZO TFT studied in this work is shown in Figure 1. First, a 200-nm-thick SiO_2 was deposited as buffer layer at 230 or 350 °C by PECVD (Applied Materials, Santa Clara, CA, USA). The F-plasma treatment (500 W, 100 mTorr, 60 s) was carried out in a reactive ion etching (RIE) chamber before depositing the IGZO. Device I is the control sample with a high-temperature buffer layer processed at 350 °C. Device II is another control sample with a low-temperature SiO_2 buffer layer processed at 230 °C. Device III is for checking the effect of F-plasma treatment on the buffer layer processed at 230 °C. A 30-nm IGZO (In: Ga: Zn = 1: 1: 1 at%) active layer (JX NIPPON MINING & METALS KOREA Co., Ltd, Pyeongtaek, Korea) was deposited by direct current sputtering at room temperature. Then, a 200-nm-thick SiO_2 layer was deposited by PECVD at 230 °C as the gate insulator (GI). The subsequent thermal annealing was performed at 230 °C for 1 h in air and followed by the sequential deposition of a 300-nm-thick Mo layer as the gate metal. The Mo and GI layers were patterned continuously by photolithography. He-plasma treatment was applied to make n$^+$ ohmic contacts with source/drain electrodes as previously reported [21,23]. A 400-nm-thick SiO_2 was deposited by PECVD at 230 °C to form the interlayer dielectric (ILD). Contact holes were formed by dry etching, after which a 300-nm-thick Mo layer was sputtered and then patterned by wet etching to form the source/drain electrode. A 200-nm-thick SiO_2 layer was deposited at 230 °C as a passivation layer by PECVD and then patterned. Finally, an indium tin oxide (ITO) was deposited and wet etched to form the pixel electrodes.

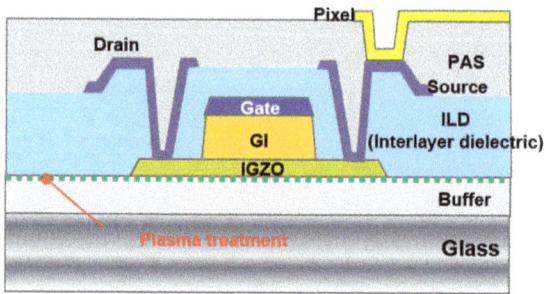

Figure 1. Cross-sectional view of the coplanar indium–gallium–zinc–oxide (IGZO) thin-film transistor (TFT) studied in this work.

3. Results and Discussions

Figure 2a shows the transfer characteristics of the devices before stress. Figure 2b–d shows the transfer characteristics of devices I, II and III before and after 1 h of PBTS, applying a constant gate voltage (V_{GS}) of +30 V at 60 °C for 1 h in a dark box. The channel width and length of the devices are 100 and 10 μm, respectively. Table 1 lists the device parameters, such as threshold voltage (V_{th}), field-effect mobility (μ_{FE}), subthreshold swing (SS) and V_{th} shift (ΔV_{th}) under PBTS for the three devices. V_{th} is the V_{GS} giving the drain current (I_D) = W/L × 10 nA at the drain voltage (V_{DS}) = 10 V. μ_{FE} is obtained at $V_{GS}-V_{th}$ = 10 V and V_{DS} = 10 V. The SS is taken from the minimum value of $\Delta V_{GS}/\Delta \log(I_{DS})$ with V_{DS} = 10 V. Devices I and II are fabricated with the same process conditions, except for the process temperature of the buffer layer. However, the mobility increases by 2.4 times and the V_{th} shift decreases from 3 to 0.2 V by using a high-temperature buffer layer. For device III, the initial V_{th}, SS, μ_{FE} and ΔV_{th} are found to be −0.2 V, 139 mV/dec, 9 cm^2 V^{-1}·s^{-1} and 0.2 V, respectively. The results show that the performance of device III is similar to that of device I, and that plasma treatment plays a critical role to improve performance.

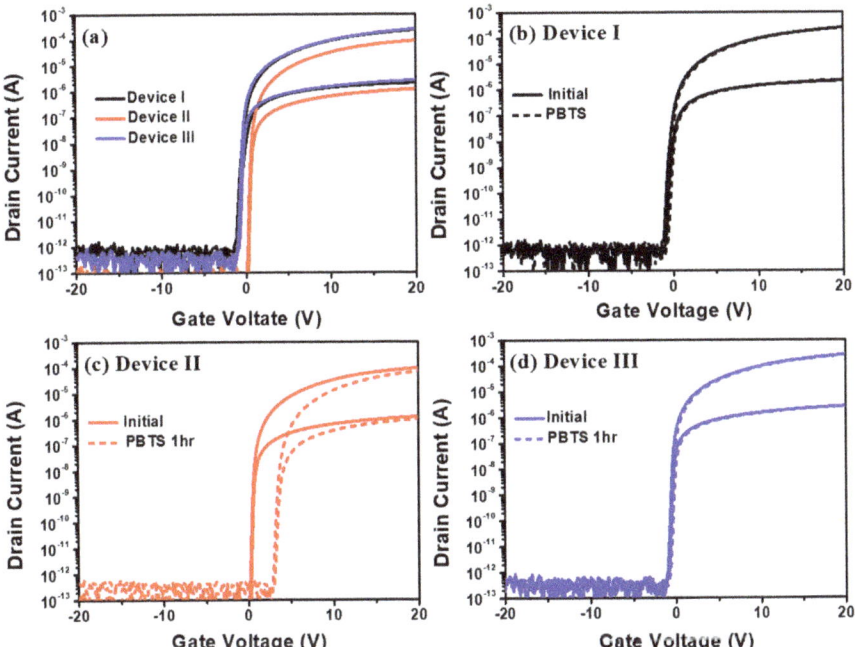

Figure 2. (a) Transfer curves of the TFTs I, II and III (W/L = 100/10 μm) before bias stress. The transfer curves were measured before and after 1 h of PBTS (dash line) at V_{DS} = 0.1 and 10 V for (b) device I, (c) device II and (d) device III, respectively.

Table 1. Summary of the device parameters and bias stability for the three TFTs. The threshold voltage shift (ΔV_{th}) was measured after 1 h PBTS.

	V_{th} (V)	μ_{FE} (cm^2/V·s)	SS (mV/dec)	ΔV_{th} (V)
Device I	0.0	9.2	181	0.2
Device II	1.1	3.8	109	3.2
Device III	−0.2	9.0	139	0.2

Figure 3a–c shows the D-SIMS profiles of H, O, F, Si and In for devices I, II and III, respectively. The significant increase of F intensity was found for device III. This result shows that F is incorporated into the IGZO/SiO$_2$ interlayer. The presence of F in the top region of the active layer of the TFT appears to be due to the contamination from the chamber wall because NF$_3$ plasma is used to etch the deposited layers to clean the PECVD chambers.

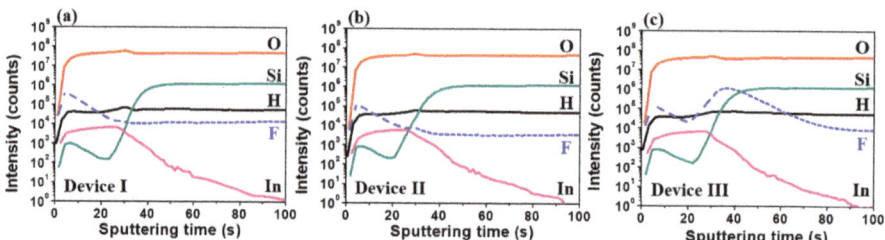

Figure 3. The dynamic secondary-ion mass spectroscopy (D-SIMS) profiles of H, O, F, Si and In for devices (**a**) I, (**b**) II and (**c**) III, respectively.

Figure 4 shows the cross-sectional HR-TEM images and annular dark-field (ADF) images in a scanning TEM (STEM) of the GI (SiO$_2$)/IGZO/buffer (SiO$_2$) interfaces. Figure 4a shows that device I has smooth surface roughness at the top and bottom regions of the active layer. This relates to a buffer layer fabricated through a high-temperature process. On the other hand, device II has poor roughness (~8 nm RMS roughness) at the bottom of the active layer, caused by the low-temperature buffer layer. Therefore, the top IGZO surface has poor surface morphology (Figure 4c). The results show the roughness of the GI/IGZO is affected by the deposition temperature of the buffer layer. Device III also has poor roughness (Figure 4f), but nanoparticles are found at the IGZO/Buffer (SiO$_2$) interface. To identify nanoparticles, the annular dark-field (ADF) images in a scanning TEM (STEM) of the devices were analyzed and the interface of device III was enlarged. Figure 4b,d,e shows ADF images for each device. ADF imaging is a method of mapping samples in a STEM. These images are formed by collecting scattered electrons with an annular dark-field detector [24]. Note that the sizes of scale bar in Figure 4d,e are larger than in Figure 4b for the identification of the nanoparticles. The nanoparticles, of around 1 nm in diameter, are identified by Figure 4e and the enlarged view of Figure 4g. The SiO$_2$ roughness affects the performance of the devices. In addition, the nanoparticles produced by plasma treatment seem to improve the mobility and V$_{th}$ shift by PBTS.

We measured the mass spectra and atomic map by atom probe tomography, APT (CAMECA, LEAP5000). APT supports the material analysis, offering extensive capabilities for both 3D imaging and chemical composition measurements at the atomic scale (around 0.1–0.3 nm resolution in depth and 0.3–0.5 nm laterally) [25]. The IGZO (100 nm)/SiO$_2$ (200 nm) layers, deposited with the same process conditions used for fabricating devices II and III, were used for APT measurement. Note that the 100-nm IGZO layer was used to prevent contamination during APT measurement. Figure 5a,b shows the mass spectra for the IGZO/buffer (SiO$_2$) interfaces of devices II and III. A 132 Da (Dalton) peak is found for the samples, which may be related with Si$_3$O$_3$ because of the atomic mass (Si: 28 Da, O: 16 Da). The peak of 134 Da is observed, which is estimated to be InF (In: 115 Da, F: 19 Da). As shown in the atomic map (Figure 5c), the InF nanoparticles of about 1 nm in diameter are formed on the top of Si$_3$O$_3$. Figure 5d shows the generation model of InF nanoparticles with the F atoms introduced by plasma treatment on the buffer layer. The InF nanoparticles could be generated during the IGZO sputtering process because In has the highest electron affinity and the largest reactivity with F. For example, the electron affinity of In is 37.043 kJ/mol, Ga is 29.061 kJ/mol and Zn is −58 kJ/mol. It is reported that F-plasma treatment on IGZO improves the TFT performance by passivating electron traps in IGZO. We found in this work that F-plasma treatment on the buffer layer formed InF nanoparticles at the IGZO/SiO$_2$ interfaces. The nanoparticles increase the density of the active layer in spite of having poor roughness. The increased density of the IGZO can improve the mobility and V_{th} shift by PBTS (Figure 2). As a result, device III has excellent TFT performance even at the low-temperature buffer-layer deposition.

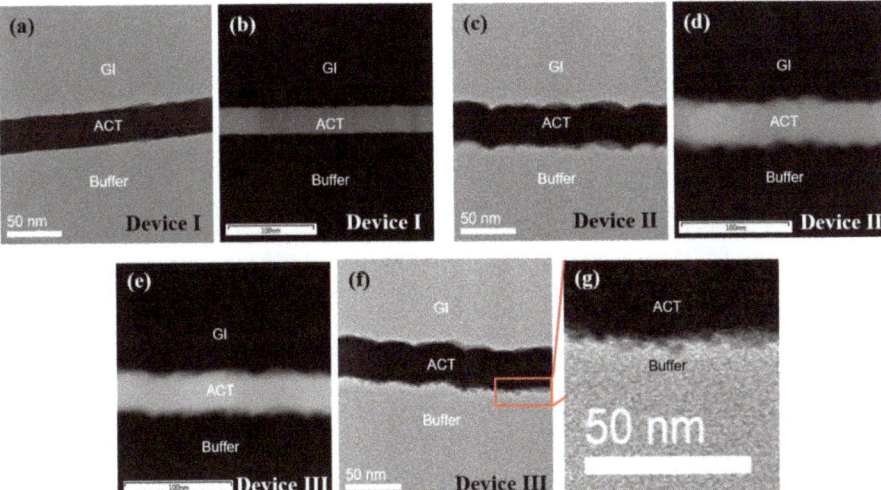

Figure 4. High-resolution TEM (HR-TEM) images of the gate insulator (GI) (SiO_2)/active (IGZO)/buffer (SiO_2) interfaces for devices (**a**) I, (**c**) II and (**f**) III, respectively. Annular dark-field (ADF) images in a scanning TEM (STEM) for devices (**b**) I, (**d**) II and (**e**) III, respectively. (**g**) Enlarged interface of device III.

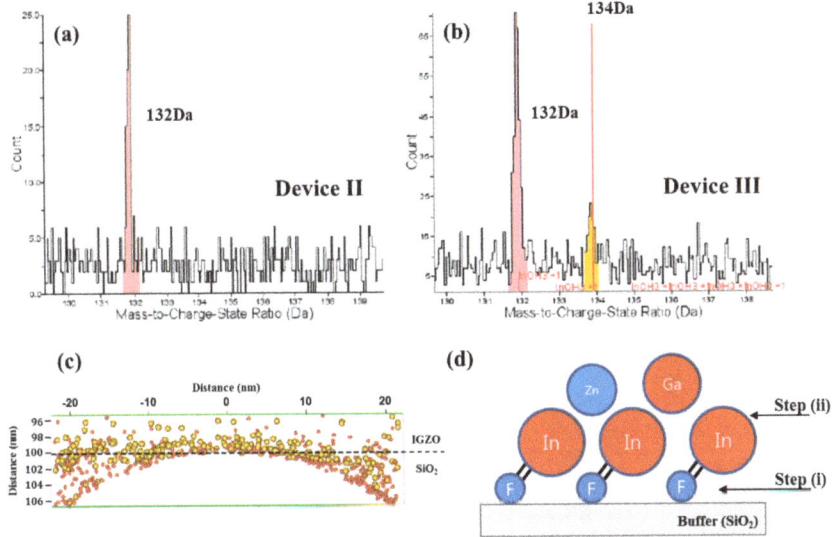

Figure 5. Mass spectra at the IGZO/buffer (SiO_2) interface for (**a**) device II and (**b**) device III. (**c**) Atomic map for device III (132 Da: red, 134 Da: yellow) analyzed with APT (atom probe tomography). (**d**) InF nanoparticle generation model. The InF formation can be seen in (**d**); step (i) is F bonding on SiO_2 and step (ii) is the formation of InF nanoparticles during the IGZO sputtering process.

4. Conclusions

We report the significant improvement of a low-temperature coplanar IGZO device by plasma exposure on an SiO_2 buffer layer. The low-temperature IGZO TFT (device III) has similar performance and V_{th} shift for PBTS to those of IGZO TFT with the buffer layer fabricated at 350 °C, by introducing F plasma on the buffer layer (230 °C). By TEM and APT analyses, it is found that In-F nanoparticles of about 1 nm in diameter are formed at the IGZO/buffer interface. Therefore, F-plasma treatment on an SiO_2 buffer layer can be a suitable method to make low-temperature coplanar IGZO TFTs.

Author Contributions: Conceptualization, H.-y.J. and J.J.; methodology, S.-h.N., K.-s.P. and S.-y.Y.; formal analysis, H.-y.J.; writing—original draft preparation, H.-y.J. and C.P.; writing—review and editing, H.-y.J.; supervision, J.J. All authors have read and agreed to the published version of the manuscript.

Funding: This work was supported by the Technology Innovation Program (or Industrial Strategic Technology Development Program (20010082, Development of low temperature patterning and heat treatment technology for light and thermal stability in soluble oxide TFT manufacturing)) funded by the Ministry of Trade, Industry & Energy (MI, Korea).

Conflicts of Interest: The authors declare no conflicts of interest.

References

1. Nomura, K.; Ohta, H.; Takagi, A.; Kamiya, T.; Hirano, M.; Hosono, H. Room-temperature fabrication of transparent flexible thin-film transistors using amorphous oxide semiconductors. *Nature* **2004**, *432*, 488–492. [CrossRef]
2. Yabuta, H.; Sano, M.; Abe, K.; Aiba, T.; Den, T.; Kumomi, H.; Nomura, K.; Kamiya, T.; Hosono, H. High-mobility thin-film transistor with amorphous InGaZnO4 channel fabricated by room temperature rf-magnetron sputtering. *Appl. Phys. Lett.* **2006**, *89*, 112123. [CrossRef]
3. Kamiya, T.; Nomura, K.; Hosono, H. Present status of amorphous In–Ga–Zn–O thin-film transistors. *Sci. Technol. Adv. Mater.* **2010**, *11*, 044305. [CrossRef] [PubMed]
4. Mativenga, M.; Choi, M.H.; Choi, J.W.; Jang, J. Transparent Flexible Circuits Based on Amorphous-Indium–Gallium–Zinc-Oxide Thin-Film Transistors. *IEEE Electron Device Lett.* **2011**, *32*, 170–172. [CrossRef]
5. Wu, C.W.; Yoo, S.Y.; Ning, C.; Yang, W.; Shang, G.L.; Wang, K.; Liu, C.H.; Liu, X.; Yuan, G.C.; Chen, J.; et al. Improvement of Stability on a-IGZO LCD. *SID Symp. Dig. Tech. Pap.* **2013**, *44*, 97–99. [CrossRef]
6. Oh, S.; Baeck, J.H.; Lee, D.; Park, T.; Shin, H.S.; Bae, J.U.; Park, K.S.; Kang, I.B. Improvement of PBTS Stability in Self-Aligned Coplanar a-IGZO TFTs. *SID Symp. Dig. Tech. Pap.* **2015**, *46*, 1143–1146. [CrossRef]
7. Nag, M.; Roose, F.D.; Myny, K.; Steudel, S.; Genoe, J.; Groeseneken, G.; Heremans, P. Characteristics improvement of top-gate self-aligned amorphous indium gallium zinc oxide thin-film transistors using a dual-gate control. *J. Soc. Inf. Disp.* **2017**, *25*, 349–355. [CrossRef]
8. Rahaman, A.; Chen, Y.; Hasan, M.M.; Jang, J. A High Performance Operational Amplifier Using Coplanar Dual Gate a-IGZO TFTs. *IEEE J. Electron Devices Soc.* **2019**, *7*, 655–661. [CrossRef]
9. Kang, D.H.; Kang, I.; Ryu, S.H.; Jang, J. Self-Aligned Coplanar a-IGZO TFTs and Application to High Circuits. *IEEE Electron Device Lett.* **2011**, *32*, 1385–1387. [CrossRef]
10. Wu, C.H.; Hsieh, H.H.; Chien, C.W.; Wu, C.C. Self-Aligned Top-Gate Coplanar In-Ga-Zn-O Thin-Film Transistors. *J. Disp. Technol.* **2009**, *5*, 515–519. [CrossRef]
11. Geng, D.; Kang, D.H.; Seok, M.J.; Mativenga, M.; Jang, J. High-Speed and Low-Voltage-Driven Shift Register with Self-Aligned Coplanar a-IGZO TFTs. *IEEE Electron Device Lett.* **2012**, *33*, 1012–1014. [CrossRef]
12. Choi, S.; Han, M. Effect of Deposition Temperature of SiOx Passivation Layer on the Electrical Performance of a-IGZO TFTs. *IEEE Electron Device Lett.* **2012**, *33*, 396–398. [CrossRef]
13. Ha, C.; Lee, H.J.; Kwon, J.W.; Seok, S.Y.; Ryoo, C.I.; Yun, K.Y.; Kim, B.C.; Shin, W.S.; Cha, S.Y. High Reliable a-IGZO TFTs with Self-Aligned Coplanar Structure for Large-Sized Ultrahigh-Definition OLED TV. *SID Symp. Dig. Tech. Pap.* **2015**, *46*, 1020–1022. [CrossRef]
14. Shin, H.J.; Takasugi, S.; Park, K.M.; Choi, S.H.; Jeong, Y.S.; Song, B.C.; Kim, H.S.; Oh, C.H.; Ahn, B.C. Novel OLED Display Technologies for Large-Size UHD OLED TVs. *SID Symp. Dig. Tech. Pap.* **2015**, *46*, 53–56. [CrossRef]

15. Yoon, J.S.; Hong, S.J.; Kim, J.H.; Kim, D.H.; Ryosuke, T.; Nam, W.J.; Song, B.C.; Kim, J.M.; Kim, P.Y.; Park, K.H.; et al. 55-inch OLED TV using Optimal Driving Method for Large-Size Panel based on InGaZnO TFTs. *SID Symp. Dig. Tech. Pap.* **2014**, *58*, 849–852. [CrossRef]
16. Arai, T. The Advantages of the Self-Aligned Top Gate Oxide TFT Technology for AM-OLED Displays. *SID Symp. Dig. Tech. Pap.* **2015**, *46*, 1016–1019. [CrossRef]
17. Ryu, M.K.; Park, S.H.K.; Hwang, C.S.; Yoon, S.M. Comparative studies on electrical bias temperature instabilities of In–Ga–Zn–O thin film transistors with different device configurations. *Solid-State Electron.* **2013**, *89*, 171–176. [CrossRef]
18. Jiang, J.; Toda, T.; Hung, M.P.; Wang, D.; Furuta, M. Highly stable fluorine-passivated In-Ga-Zn-O thin-film transistors under positive gate bias and temperature stress. *Appl. Phys. Express* **2014**, *7*, 114103. [CrossRef]
19. Jiang, J.; Tatsuya, T.; Tatsuoka, G.; Wang, D.; Furuta, M. Improvement of Electrical Properties and Bias Stability of InGaZnO Thin-Film Transistors by Fluorinated Silicon Nitride Passivation. *{ECS} Trans.* **2014**, *64*, 59–64. [CrossRef]
20. Yamazaki, H.; Ishikawa, Y.; Fujii, M.; Ueoka, Y.; Fujiwara, M.; Takahashi, E.; Andoh, Y.; Maejima, N.; Matsui, H.; Matsui, F.; et al. The Influence of Fluorinated Silicon Nitride Gate Insulator on Positive Bias Stability toward Highly Reliable Amorphous InGaZnO Thin-Film Transistors. *{ECS} J. Solid State Sci. Technol.* **2013**, *3*, 20–23. [CrossRef]
21. Um, J.G.; Jang, J. Heavily doped n-type a-IGZO by F plasma treatment and its thermal stability up to 600 °C. *Appl. Phys. Lett.* **2018**, *112*, 162104. [CrossRef]
22. Lee, S.; Shin, J.; Jang, J. Top Interface Engineering of Flexible Oxide Thin-Film Transistors by Splitting Active Layer. *Adv. Funct. Mater.* **2017**, *27*, 1604921. [CrossRef]
23. Jeong, H.Y.; Lee, B.Y.; Lee, Y.J.; Lee, J.I.; Yang, M.S.; Kang, I.B.; Mativenga, M.; Jang, J. Coplanar amorphous-indium-gallium-zinc-oxide thin film transistor with He plasma treated heavily doped layer. *Appl. Phys. Lett.* **2014**, *104*, 022115. [CrossRef]
24. Otten, M.T. High-Angle annular dark-field imaging on a tem/stem system. *J. Electron Microsc. Tech.* **1991**, *17*, 221–230. [CrossRef]
25. Herbig, M.; Choi, P.; Raabe, D. Combining structural and chemical information at the nanometer scale by correlative transmission electron microscopy and atom probe tomography. *Ultramicroscopy* **2015**, *153*, 32–39. [CrossRef]

© 2020 by the authors. Licensee MDPI, Basel, Switzerland. This article is an open access article distributed under the terms and conditions of the Creative Commons Attribution (CC BY) license (http://creativecommons.org/licenses/by/4.0/).

Article

Effects of Annealing on Characteristics of Cu₂ZnSnSe₄/CH₃NH₃PbI₃/ZnS/IZO Nanostructures for Enhanced Photovoltaic Solar Cells

Chzu-Chiang Tseng [1,2], Gwomei Wu [1,2,*], Liann-Be Chang [1], Ming-Jer Jeng [1], Wu-Shiung Feng [1], Dave W. Chen [2], Lung-Chien Chen [3] and Kuan-Lin Lee [3]

1. Institute of Electro-Optical Engineering, Department of Electronic Engineering, Chang Gung University, Taoyuan 333, Taiwan; D0427101@cgu.edu.tw (C.-C.T.); liann@mail.cgu.edu.tw (L.-B.C.); mjjeng@mail.cgu.edu.tw (M.-J.J.); fengws@mail.cgu.edu.tw (W.-S.F.)
2. Chang Gung Memorial Hospital, Keelung 204, Taiwan; mr5181@cgmh.org.tw
3. Department of Electro-Optical Engineering, National Taipei University of Technology, Taipei 106, Taiwan; Ocean@ntut.edu.tw (L.-C.C.); t102658016@gmail.com (K.-L.L.)
* Correspondence: wu@mail.cgu.edu.tw; Tel.: +886-3-211-8800

Received: 30 January 2020; Accepted: 10 March 2020; Published: 13 March 2020

Abstract: This paper presents new photovoltaic solar cells with Cu₂ZnSnSe₄/CH₃NH₃PbI₃(MAPbI₃)/ZnS/IZO/Ag nanostructures on bi-layer Mo/FTO (fluorine-doped tin oxide) glasssubstrates. The hole-transporting layer, active absorber layer, electron-transporting layer, transparent-conductive oxide layer, and top electrode-metal contact layer, were made of Cu₂ZnSnSe₄, MAPbI₃ perovskite, zincsulfide, indium-doped zinc oxide, and silver, respectively. The active absorber MAPbI₃ perovskite film was deposited on Cu₂ZnSnSe₄ hole-transporting layer that has been annealed at different temperatures. TheseCu₂ZnSnSe₄ filmsexhibitedthe morphology with increased crystal grain sizesand reduced pinholes, following the increased annealing temperature. When the perovskitefilm thickness was designed at 700 nm, the Cu₂ZnSnSe₄ hole-transporting layer was 160 nm, and the IZO (indium-zinc oxide) at 100 nm, and annealed at 650 °C, the experimental results showed significant improvements in the solar cell characteristics. The open-circuit voltage was increased to 1.1 V, the short-circuit current was improved to 20.8 mA/cm², and the device fill factor was elevated to 76.3%. In addition, the device power-conversion efficiency has been improved to 17.4%. The output power P_{max} was as good as 1.74 mW and the device series-resistance was 17.1 Ω.

Keywords: CZTSe; hole-transporting material; perovskite; IZO; magnetron sputtering

1. Introduction

Photovoltaic (PV) devices provide electrical energy directly from sunlight, and have been one of the promising technologies in the renewable energy industry. The further improvement in the efficiency and reliability and also reduction in cost are in great demand, for the healthy development of global economy. The organic metal halide materials introduced a new generation ofthin-filmPVs, due to the excellent characteristics for light harvesting in solar cells. The optical-to-electrical power-conversion efficiency (PCE) of the lead halide perovskite- (CH₃NH₃PbI₃, or MAPbI₃) based PV device has been increased substantially in recent years [1–5]. The MAPbI₃ substance favors efficient carrier generation and transport to electrodes. It can absorb sunlight radiation from ultra-violet to infrared region. The electron-hole diffusion length could exceed 1 µm in a tri-halide perovskite absorber [6]. Kim et al. developed the first solid-state lead halide perovskite PV with fluorinated tin oxide (FTO)/TiO₂/MAPbI₃/2,2′,7,7′–tetrakis(*N*,*N*-di-*p*-methoxyphenylamino)-,9,9′ spiro-bifluorene (SpiroOMeTAD)/Au nanostructures [7]. Later, Jeng et al. reported the first inverted planar structure of a lead halide perovskite solar cell [8]. Choi et al. studied conjugated polyelectrolytes as the hole-transporting materials (HTM) for inverted-type perovskite solar cells [9]. However, the organic HTM could corrode the electrode materials, which would, in turn, reduce the cell stability. Thus, inorganic type materials, such as CdTe, CuZnSnS, CuZnSnSe, CuZnSnSSe, CuInGaS, CuInGaSe, and CuInGaSSe have become interesting topics for the perovskite solar cells [10–13]. In addition, FTO should be used in favor overindium-tin-oxide (ITO) as the transparent conducting oxide(TCO), when high-temperature annealing in air is in need for PV device fabrication. This is because the ITO electrical properties can be degraded in the presence of oxygen at relatively high temperature. FTO is more stable after such a high-temperature annealing process. This good transparent conductive layer could significantly decrease photon absorption at the back electrode.

Li et al. reviewed the architectures and deposition methods in standard formation of perovskite and charge-transport layers, and provided an overview on PV device stability [14]. Christians et al. developed tailored interfaces to increase the operational stability of un-encapsulated perovskite solar cells [15]. The long-term stability performance has to be improved for different ingredients to warrant inter-carrier scalability, capability, and system durability. The actual PV devices involved frames, glasses, and other compound materials, which could further challenge the recycling scheme [16]. In addition, metal-selenide had been studied as a carrier-transporting material in perovskite by various sputtering techniques [17–19]. This paper has focused on applying$Cu_2ZnSnSe_4$as HTM nano-film, and fabricated by innovative concept to improve Cu-based PV performance. The HTM film should help to maintain perovskite's open-circuit voltage (Voc), short-circuit current density (J_{SC}), fill factor (FF), and stability. An ultra-thin CuZnSnSe HTM (<200 nm) between Mo metal-electrode layer and $MAPbI_3$ active absorber layer, would promote carrier transporting, and provide favorable ohmic-contact. The deposition of a good ohmic-contact can reduce interface carrier recombination.

Thin CuInGeSe or CuZnSnS layer could fail to maintain the high efficiency performance. The main reason behind the poor performance was the substantial drop in short-circuit current density [20–22]. The Kesterite photovoltaics that applied for Cu_2ZnSnS_4, $Cu_2ZnSnSe_4$, and $Cu_2ZnSn(S_{1-x}Se_x)_4$ emerged as one of the most assured replacements for the chalcopyrite solar cells. The constituent elements are relatively abundant on the earth. They have also demonstrated high absorption coefficients, and direct tunable energy band gaps that would allow effective absorption of photons [23]. The $Cu_2ZnSnSe_4$ HTM film can be deposited in 40~160 nm thickness by magnetron sputtering to achieve well carrier transport. The inorganic CuZnSnSe HTM provides low-cost procedure and material for the applications of perovskite solar cells.

Chalapathy et al. revealed high-quality CZTS films by high-temperature sulfurization of sputtered Cu/ZnSn/Cu precursor layers [24]. A highly imperfect $Cu_2ZnSnSe_4$ HTM/TCO material interface could be an exhaustion region, leading to high back-electrode contact layer of surface recombination. The selenium diffuses into Mo film's interface, resulting in possible selenization. The $MoSe_2$ film at the interface indicated structural quality for adherence and good electrical contact [25]. The outward diffusion of selenium into Mo strongly depends on the annealing temperature [26]. The reconstructed interfacial layer significantly diminishes series-resistance and enlarges shunt-resistance of the solar cell.

Aqil et al. studied the electrical and morphological appearance of the functional portion of Mo film deposition [27]. A single layer of Mo film with low resistivity and good adhesion has been deposited successfully by the radio-frequency (RF) magnetron sputtering system using appropriate parameters. The nanocrystals could be tuned by the nucleation and growth conditions [28]. The bi-layer Mo films were thus proposed to consist of a porous bottom layer and a dense top layer [29]. It has been possible to control the interfacial $MoSe_2$ film thickness during annealing procedure to achieve high solar cell power conversion efficiency [30,31]. The interface of CIGSe/Mo or CZTSe/Mo could improve the structural quality of planar solar cells, and affects the adherence to metal-contacts, providing excellent ohmic-contact at CIGSe/Mo or CZTSe/Mo inter-films. The development of $MoSe_2$ could not be formed below 500 °C. The selenization increased strongly on higher annealing temperature, also leading to nano-crystallization orientation. Nevertheless, theCIGSe/Mo or CZTSe/Mo heterojunction could be associated with MoSe$_2$ film. It was not an interrelated Schottky-type but was an interconnected ohmic-contact [32,33].

Furthermore, ZnS is an ideal inorganic compound to be used as an electron-transporting layer (ETL)for the perovskite PVs [34]. When ZnS dense film is synthesized, it can be transparent, and can be used as a window for visible optics and infrared optics. The film's nanostructure relies on the deposition procedure or doping element at the growth sequence. In comparison with other complex techniques, this study employed RF (radio frequency) magnetron sputtering process for stable depositing results [35]. The MAPbI$_3$/ZnS interface property has been demonstrated by using ultra-violet photoelectron spectra [36]. The ZnS interfacial nano-film could promote the Voc and perovskite PCE performance on PVs. The cascade conduction band architecture effectively decreased the interfacial charge recombination and improved the electron transfer. In addition, Yuan et al. illustrated the reduction in surface defects, which improved charge extraction, extended light response, and expanded ZnS/MAPbI$_3$ spectral absorption [37].

Indium-doped zinc oxide (IZO) film only needs a low annealing temperature and can promote the optoelectronic characteristics. It is an amorphous transparent conducting oxide material, and has a downgraded absorption near IR (infra red) region photons, yielding to an enhanced transmission at the bottom layer on solar cells. The IZO material is a potential replacement of traditional TCO that was used as an n-type buffer layer or window transparent conductive oxide layer for photovoltaics. Nevertheless, the fabrication using all sputtering-process methodology, along with the ultra-thin Cu$_2$ZnSnSe$_4$ HTM, may develop a novel copper-based inorganic HTM for perovskite nanostructured PVs. The results should help to pave the way for the next generation thin-film solar cells and to better protect the global environment.

2. Materials and Methods

In this study, bi-layer Mo film was sputtered on FTO glass substrate as a back metal electrode contact layer. The Mo film was prepared by RF magnetron sputtering system using commercial Mo target (Ultimate Materials Technology Co., Miaoli, Taiwan). In this fabrication process, the bottom layer was deposited at a higher Ar flow working pressure, using high-power parameters, and the Ar flow rate and RF power were maintained at 70 sccm and 110 W. On the other hand, the top layer was deposited at a lower Ar flow working pressure, using lower-power parameters, and the Ar flow rate and RF power were maintained at 35 sccm and 55 W for preserving better adhesion. The bi-layer Mo films exhibited both low resistivity and good adhesion. It has been measured that the Mo bi-layer had a film resistivity of 4.37×10^{-4} Ω-cm, which was much lower than that of single-layer at 2.2×10^{-2} Ω-cm. Each layer had a thickness of ~100 nm. The Mo film also acted as a reflective layer on these multi-layered solar cells. This Mo film that was deposited under high argon pressure would be under tensile stress and adhere successfully with the substrate, but with low conductivity. The deposition by low argon pressure would render compressive stress that had low resistivity but adhered poorly to interface on the substrate.

Moreover, Cu$_2$ZnSnSe$_4$ film was deposited by RF magnetron sputtering using a Cu$_2$ZnSnSe$_4$ target (Ultimate Materials Technology Co., Miaoli, Taiwan) on bi-layer Mo. The argon flow rate and RF power were maintained at 40 sccm and 70 W, respectively. It was adopted as the HTM layer. It should promote the carrier transporting and result in a beneficial device by providing a conductive ohmic-contact. The ultra-thin Cu$_2$ZnSnSe$_4$ HTM (<200 nm) surface roughness corresponded to the optical absorber thickness transition, and could affect interface recombination of electrons and electron-holes. The root-mean-square surface roughness was low at ~20 nm, measured by atomic force microscopy. The Cu$_2$ZnSnSe$_4$ film thickness has been prepared at 40–160 nm approximately. It was then further thermally treated by the annealing temperature at 350, 450, 550, or 650 °C in a tube furnace for about 60 min in order to get magnificent crystallization.

The MAPbI$_3$ film was deposited on grown Cu$_2$ZnSnSe$_4$ HTM layer and by one-step spin-coating process for the inverted structures of perovskite solar cells. The photovoltaic characteristics would be investigated to reveal the relationships between the properties and structures. The single-step deposition involved the dissolution of PbI$_2$ and MAI (CH$_3$NH$_3$I) in a co-solvent, consisted of equal volumes of dimethyl sulfoxide and gamma-butyrolactone. This perovskite precursor solution was spin-coated using parameters of 1000 and 5000 rpm for 10 and 18 s, respectively, in a nitrogen-filled glove box. The wet film was then quenched by dropping 50 µL of anhydrous toluene at 15 s. Afterwards, the perovskite film was further annealed at 100 °C for 10 min. The MAPbI$_3$ thin-film had a thickness of 700 nm approximately.

Zinc sulfide film is an *n*-type semiconductor material and was adopted as an ETL in this multi-layered nanostructure PVs. It was prepared using a commercial target (Ultimate Materials Technology Co., Miaoli, Taiwan) by RF sputtering system. The Ar flow rate and RF power were controlled at 30 sccm and 50 W, respectively. The ZnS film was about 50 nm in thickness. The film was attributed to not only thinner ETL layer deposition, but also to extract electrons from the MAPbI$_3$ active absorber layer. It also required quenching at 100 °C in a tube furnace for about 10 min to achieve the ideal *p-n* junction, crystallization, and better ohmic-contact. The ZnS film could improve the multi-layer structures by alloying, plastic distortion and thermal annealing. It would bond with upper IZO transparent conductive oxide film, while conducting the necessary optical-current passing through, thus enhancing optical-to-electrical conversion performance.

Indium-doped zinc oxide is an n-type semiconductor material and was adopted as a transparent conductive oxide film, deposited by RF magnetron sputtering using a commercially available IZO target (Ultimate Materials Technology Co., Miaoli, Taiwan). The deposition parameters were argon flow rate at 30 sccm and RF power at 50 W. The IZO film had a thickness of 100 nm approximately. The low-temperature and high-mobility amorphous nature rendered excellent characteristics for the ZnS (ETL)/MAPbI$_3$/CZTSe (HTM) heterojunction solar cells.

At last, the top Ag metal electrode film was deposited over the IZO *n*-type semiconductor TCO. The Ag metal ingot was prepared on a tungsten metal-boat in the vacuum chamber of an evaporation system. The tungsten boat was connected to an external power supply and provided with a maximum current of 90 A. The Ag metal film had a thickness of ~100 nm. A shadow mask has been adopted to define an active area of 0.5×0.2 cm^2 during the Ag deposition. The nanostructured PV was investigated by X-ray diffraction (XRD) using PANalytical X'Pert Pro DY2840 system (Malvern Panalytical, Almelo, Netherlands) with Cu Kα (λ= 0.1541 nm) radiation. The crystalline surface morphology was studied by scanning electron microscopy (Zeiss Gemini SEM, Jena, Germany). A micro-Raman spectroscopy analysis was employed using Jon-YvonLabRAM system (Horiba-HR800, Kyoto, Japan). The photoluminescence (PL) results were scanned by a fluorescence spectrophotometer (Hitachi, F-7000, Tokyo, Japan). The electron spectroscopy chemical analysis (ESCA) spectra were examined using PHI-5000 system (ULVAC, Versaprobe-II, Kanagawa, Japan). The solar cell PV characteristics were studied by a Keithley 2420 programmable source instrument under 1000 W xenon illumination with a forward scan rate of 0.1 V/s.

3. Results and Discussion

Figure 1 shows the complete scheme of the nanostructured MAPbI$_3$ perovskite solar cell device with the Cu$_2$ZnSnSe$_4$ HTM layer. The corresponding energy levels of the planar architecture device of Ag/IZO/ZnS/MAPbI$_3$/Cu$_2$ZnSnSe$_4$/Mo/FTO are illustrated in Figure 2. The ultra-thin Cu$_2$ZnSnSe$_4$ HTM has been deposited between the bi-layer Mo metal-electrode and the MAPbI$_3$ active absorber layer to improve carrier transporting. The energy level diagram indicated that it is a heterojunction planar photovoltaic solar cell.

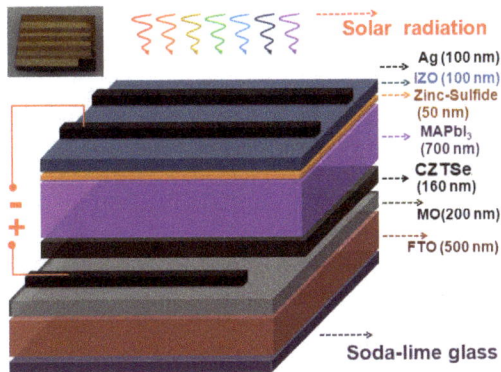

Figure 1. The complete scheme of nanostructured MAPbI$_3$ perovskite solar cell with Cu$_2$ZnSnSe$_4$ HTM layer.

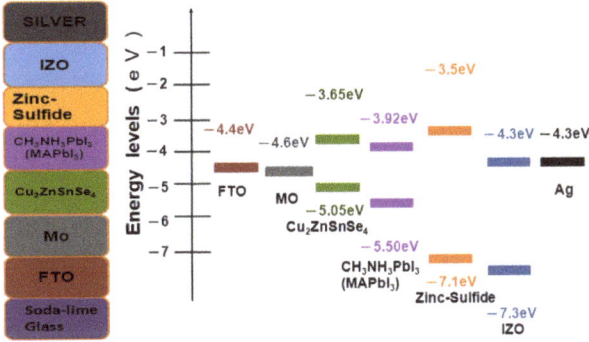

Figure 2. The corresponding energy levels of the planar architecture device of Ag/IZO/ZnS/MAPbI$_3$/Cu$_2$ZnSnSe$_4$/Mo/FTO.

Figure 3 shows the XRD diffraction pattern results of the MAPbI$_3$/Cu$_2$ZnSnSe$_4$/Mo/FTO nanostructures on glass substrate after the various annealing temperatures [3,21,31]. The MAPbI$_3$ film's nano-crystal was illustrated by one main crystal plane (110) corresponding to the 2θ diffraction peak at ~14.3°. The 2θ full-width at half-maximum (FWHM) was reduced from 0.39° to 0.28° when the annealing temperature was increased from 350 °C to 650 °C. Other MAPbI$_3$ crystal planes involved (220) at ~29.2° and (310) at ~32.4°. When the annealing temperature of the Cu$_2$ZnSnSe$_4$ HTM film under the MAPbI$_3$ film was increased, the FWHM of the Cu$_2$ZnSnSe$_4$ HTM film's nano-crystal was also improved. Its nano-crystal was illustrated by the main crystal plane (112) corresponding to the 2θ diffraction peak at ~27.1°. Its FWHM was reduced from 0.64° to 0.51° when the annealing temperature was increased from 350 to 650 °C. Other Cu$_2$ZnSnSe$_4$ crystal planes could be illustrated by (204) at 2θ diffraction peak of ~45.1°, (312) at ~53.8°, and (008) at 65.8°. The Mo film's nano-crystal was illustrated by the main crystal plane (110) at the 2θ diffraction peak at ~40.5°. Its FWHM was reduced from 0.98° to 0.83° when the annealing temperature was increased from 350 to 650 °C. Other Mo film's crystal planes included (200) at ~58.6°, and (211) at ~73.5°. The smaller FWHM demonstrated MoSe$_2$ nano-crystal and it could be illustrated by the main crystal plane (002) at the 2θ diffraction peak at ~13.5°. The FWHM was reduced from 0.13° to 0.10° when the annealing temperature was increased from 350 to 650 °C. Other MoSe$_2$ crystal planes included (004) at ~28.2°, (006) at ~42.7°, and (008) at ~57.5°. The Cu$_2$ZnSnSe$_4$ HTM film has been annealed at 350, 450, 550, and 650 °C, respectively. As a result, the film's crystal quality was improved following the increased annealing temperature. Interestingly, the crystallization of MAPbI3 active absorber layer was also preceded by the same annealing temperature. It has been noted that when one side of a heterojunction is much more heavily doped than the other side, the junction is nearly a one-sided heterojunction. To form a good quality heterojunction, the difference between the neighboring semiconductors' lattice constants should be small in order to minimize the density of interface states. The difference in electron hole affinity between two different materials should be small to minimize band discontinuity, and thermal expansion coefficients should be close as well. MAPbI3 film and Cu2ZnSnSe4 HTM film were likely annealed and sintered altogether.

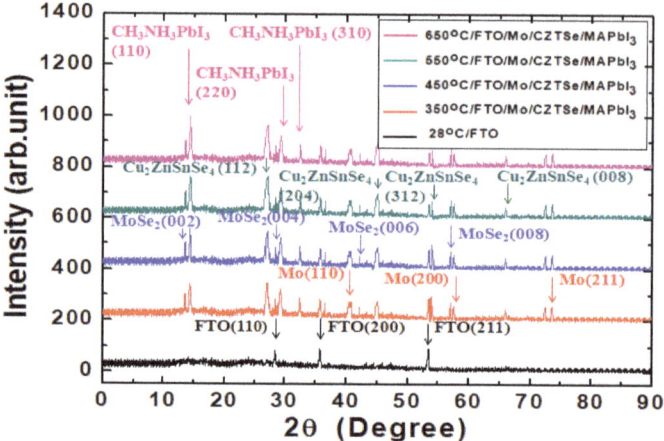

Figure 3. XRD diffraction pattern results of the MAPbI3/Cu2ZnSnSe4/Mo/FTO nanostructures on glass substrate after the various annealing temperatures.

Furthermore, their energy band gaps were similar for trapping light simultaneously. It would be beneficial to constitute pairs of excited electrons and associated electron holes. Eventually, the carriers could increase the optical-electronic power-conversion efficiency and optical-current associated in the photovoltaic cell's multi-layer nanostructures.

Figure 4 shows the top-view SEM micrographs of the surface morphology of Cu$_2$ZnSnSe$_4$ HTM layers after the various annealing temperatures. The magnetron sputtered film provided full surface coverage and was composed of small crystal grains ranging from tens of nm to one μm in size. After its deposition, the bi-layer Mo film became indiscernible. However, the Cu$_2$ZnSnSe$_4$ nano-crystal grainsize was significantly enlarged with the increased annealing temperature. For the 350 °C-annealed sample, it exhibited relatively small crystal grains from tens to two hundred nm. It contained some pinhole-type of defects. Figure 4b showed Cu$_2$ZnSnSe$_4$ film surface appearance with tens to four hundred nm in grain sizefor 450 °C. Figure 4c showed film surface with tens to six hundred nm in the grain sizefor 550 °C. It also contained less pinholes at the same magnification. Figure 4d showed the Cu$_2$ZnSnSe$_4$ film surface morphology for 650 °C. It demonstrated better crystallization, with crystal grains as large as one μm in size. Much less pinholes could be found. Thus, the higher annealing temperature achieved high-quality surface HTM films. Additionally, the Cu$_2$ZnSnSe$_4$ HTM film's hole mobility was increased from 15.1 cm^2/(V s) to 29 cm^2/(V s), while the annealing temperature was increased from 350 to 650 °C. The higher carrier mobility would help to reduce the device series-resistance, and improve performance for the nanostructured photovoltaic cells.

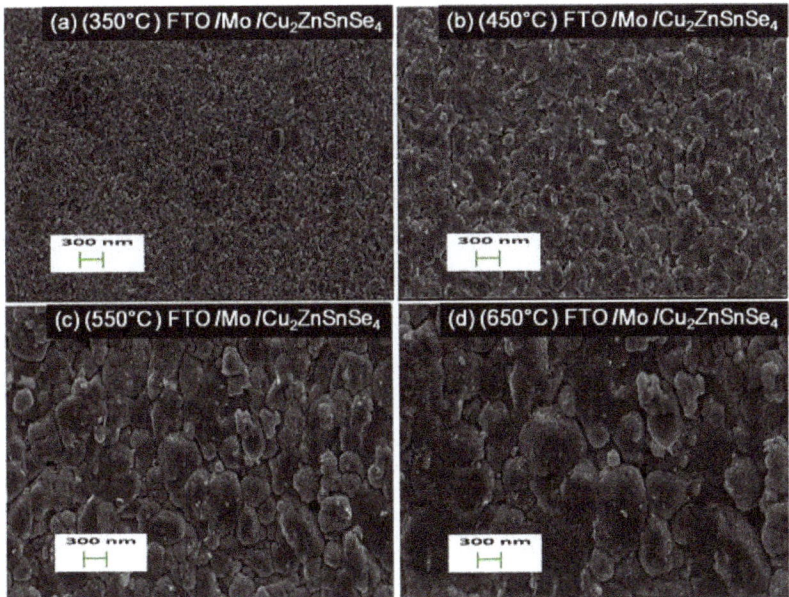

Figure 4. Top-view SEM micrographs of theCu$_2$ZnSnSe$_4$ HTM layers after the various annealing temperatures of: (**a**) 350 °C; (**b**) 450 °C; (**c**) 550 °C; (**d**) 650 °C.

Figure 5 shows the compositional dependence of Raman spectra of the Cu$_2$ZnSnSe$_4$ HTM nano-films after the various annealing temperature treatments. In this fabrication, Cu$_2$ZnSnSe$_4$ HTM film exhibited dominant spectra with intense Raman scattering main peak at 194 cm^{-1} correlating with the optical phonon mode. Its intensity increased slightly with the increased thermal annealing temperature. Other Raman scattering peak intensities were found at 233 and 253 cm^{-1}. The Cu$_2$ZnSnSe$_4$ HTM crystal orientation could be found from polarization of Raman-scattered light with respect to the laser light, if the crystal structure's point group could be known.

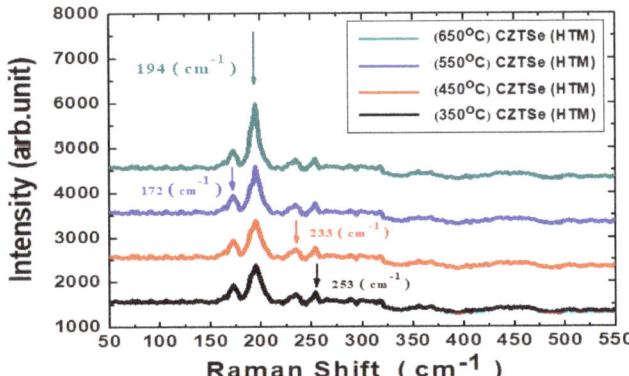

Figure 5. The compositional dependence of Raman spectra of the Cu$_2$ZnSnSe$_4$ HTM nano-films. The arrows point towards the corresponding wave numbers.

Figure 6 shows the measurement results of the absorbance spectra of Cu$_2$ZnSnSe$_4$ HTM nano-films after the various annealing temperatures. The optical absorption properties of Cu$_2$ZnSnSe$_4$ in the visible region and near-infrared region are associated with electronic transitions and are also useful in comprehending electronic band conformations of semiconducting films. The optical spectra were recorded using a UV (ultraviolet) spectrophotometer at the wavelength range from 300 to 1200 nm. It was observed that the absorbance intensity increased with the increased annealing temperature. This was presumably caused by free-carrier absorption corresponding to conductivity. These absorption spectra illustrated that all $Cu_2ZnSnSe_4$ nano-films absorb over the entire visible region of electromagnetic waves. The absorption spectra data were analyzed following a classical equation for near edge optical absorption of semiconductors: $\alpha h\nu = A(h\nu - E_g)^n$, where α is absorption coefficient, $h\nu$ is photon energy, E_g is energy band gap, A is constant, n can have values of 1/2, 2, 3/2 and 3 for allowed direct, allowed indirect, forbidden direct and forbidden indirect transitions, separately. The $Cu_2ZnSnSe_4$ energy band gap was determined by plotting a graph of $h\nu$ versus $(\alpha h\nu)^2$, for the direct band gap. The energy band gap was designated by extrapolating the straight line portion to the energy axis, whose intercept to the x-axis should give the optical energy band gap [38]. An example graph for $Cu_2ZnSnSe_4$ annealed at 650 °C is provided in Figure 7, and the calculation result of energy band gap has been 1.07–1.1 eV for the ultra-thin $Cu_2ZnSnSe_4$ HTM which was deposited on bi-layer Mo/FTO glass substrate. It should promote a valence electron bound to an atom to a conduction electron. Such electrons then move freely within the HTM and become carriers that can conduct current.

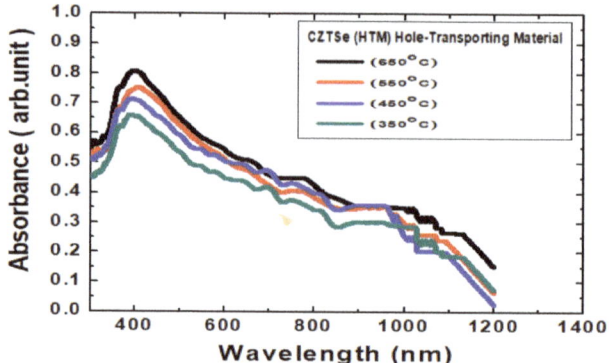

Figure 6. The absorbance spectra of $Cu_2ZnSnSe_4$ HTM nano-films after the various annealing temperatures.

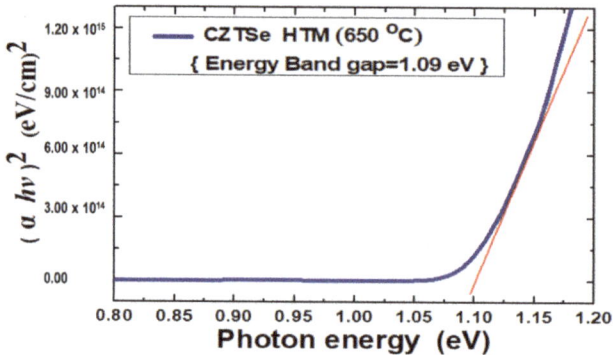

Figure 7. Graph of $h\nu$ versus $(\alpha h\nu)^2$ for measuring the optical energy bandgap of $Cu_2ZnSnSe_4$. It was determined by extrapolating the straight line portion to the photon energy axis.

Figure 8 displays the graphic current-density voltage (J–V) curves of Ag/IZO/ZnS/Cu$_2$ZnSnSe$_4$/Mo/FTO nanostructured solar cells. The Cu$_2$ZnSnSe$_4$ HTM layer thickness has been varied at 40~160 nm, and the thermal annealing temperature was all at 650 °C. Additionally, Table 1 summarizes the PV characteristic parameters of these nanostructured solar cells, without MAPbI$_3$ perovskite, under 100 mW/cm^2 illumination (air mass, AM1.5G). It has been evidenced that the open-circuit voltage increased from 0.36 to 0.39 V, following the increased HTM layer thickness from 40 to 160 nm. The device short-circuit current was also enlarged from 6.47 to 9.46 mA/cm^2. The device fill factor value would be amplified from 39.5% to 46.3%. The PV device power-conversion efficiency value was slightly increased from 0.92% to 1.71%, and the output power P_{max} value was enhanced from 0.09 to 0.17 mW. Additionally, the Cu$_2$ZnSnSe$_4$ film alone could not absorb enough photons, so that the PV cells exhibited poor PCE performance in the illustration.

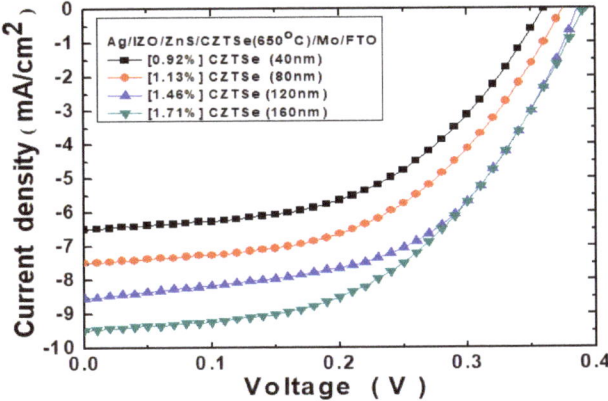

Figure 8. The J–V curves of Ag/IZO/ZnS/Cu$_2$ZnSnSe$_4$/Mo/FTO nanostructured solar cells, under 100 mW/cm^2 illumination. The Cu$_2$ZnSnSe$_4$ HTM layer thickness has been varied at 40–160 nm.

Table 1. The PV characteristic parameters of the Ag/IZO/ZnS/Cu$_2$ZnSnSe$_4$/Mo/FTO nanostructured solar cells.

Cu$_2$ZnSnSe$_4$ HTM Thickness (nm)	V_{OC} (V)	J_{SC} (mA/cm^2)	FF (%)	Eff (%)	P_{max} (mW)
40	0.36	6.47	39.5	0.92	0.09
80	0.37	7.47	40.9	1.13	0.11
120	0.38	8.52	45.1	1.46	0.15
160	0.39	9.46	46.3	1.71	0.17

Figure 9 displays the graphic J–V curves of Ag/ZnS/MAPbI$_3$/Cu$_2$ZnSnSe$_4$/Mo/FTO nanostructured solar cells, with the MAPbI$_3$ perovskite nanostructures on bi-layer Mo. The Cu$_2$ZnSnSe$_4$ HTM layer thickness has been fixed at 160 nm, but was thermally treated at the various annealing temperature 350–650 °C. The measurements were, again, under 100 mW/cm^2 illumination. Table 2 lists the derived PV characteristic parameters of the Ag/ZnS/MAPbI$_3$/Cu$_2$ZnSnSe$_4$/Mo/FTO nanostructured solar cells. It has been clearly evidenced with significant improvements in all the parameters. The open-circuit voltage was increased from 0.89 to 0.98 V following the increased annealing temperature from 350 to 650 °C. The short-circuit current was also expanded from 19.9 to 20.4 mA/cm^2. The device fill factor value could be increased from 68.6% to 71.3%. The PV device power-conversion efficiency value was strengthened from 12.2% to 14.3%, and the output power P_{max} value was enhanced from 1.22 to 1.43 mW.

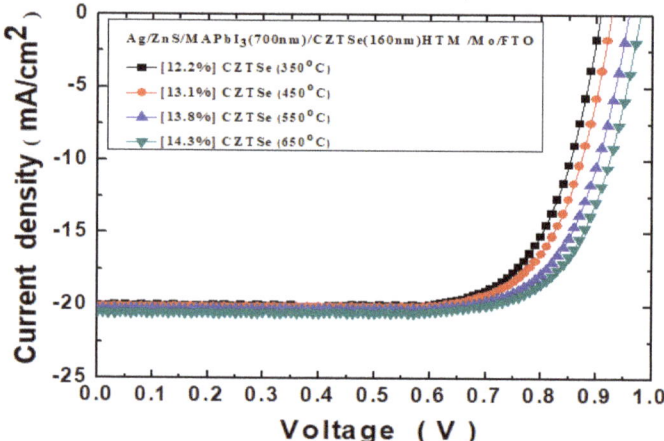

Figure 9. The J–V curves of Ag/ZnS/MAPbI$_3$/Cu$_2$ZnSnSe$_4$/Mo/FTO nanostructured solar cells, under 100 mW/cm^2 illumination. The Cu$_2$ZnSnSe$_4$ HTM has been thermally treated at the various annealing temperature 350–650 °C.

Table 2. The PV characteristic parameters of the Ag/ZnS/MAPbI$_3$/Cu$_2$ZnSnSe$_4$/Mo/FTO nanostructured solar cells.

Cu$_2$ZnSnSe$_4$ HTM Annealing Temperature (°C)	V_{OC} (V)	J_{SC} (mA/cm^2)	FF (%)	Eff (%)	P_{max} (mW)
350	0.89	19.9	68.6	12.2	1.22
450	0.92	20.1	70.9	13.1	1.31
550	0.95	20.2	71.2	13.8	1.38
650	0.98	20.4	71.3	14.3	1.43

In addition, Figure 10 displays the J–V curves of the Ag/IZO/ZnS/MAPbI$_3$/Cu$_2$ZnSnSe$_4$/Mo/FTO nanostructured solar cells at the various HTM thermal annealing temperatures under 100 mW/cm^2 illumination. The TCO layer in thickness of 100 nm IZO film has been inserted between the ZnS ETL and top Ag electrode. Table 3 presents the derived PV characteristic parameters of the Ag/IZO/ZnS/MAPbI$_3$/Cu$_2$ZnSnSe$_4$/Mo/FTO nanostructured solar cells. Furthermore, it has been clearly evidenced with enhancements in all the parameters. The open-circuit voltage was amplified from 0.97 to 1.10 V, following the annealing temperature that was increased from 350 to 650 °C. The short-circuit current was slightly increased from 20.5 to 20.8 mA/cm^2. The device fill factor value was slightly diminished to 76.3%. The PV device power-conversion efficiency value was further increased from 15.5% to 17.4%. Additionally, the device series-resistance was decreased from 20.2 to 17.1 Ω., and the device output power P_{max} value could be enhanced from 1.55 to 1.74 mW.

Figure 11 shows the PL spectral measurement results of MAPbI$_3$ perovskite nano-films onCu$_2$ZnSnSe$_4$/Mo/FTO following the various thermal annealing temperatures. The spectra were examined by fluorescence spectrophotometer. It has been clearly evidenced with one main peak at the wavelength of ~768 nm. The intensity was also enhanced by the increased annealing temperature. The intensity of PL spectrum is relative to lifetime of the injected electrons and electron-holes combined to form excitons. An exciton indicates a mobile energy constitution by an excited electron and a relative electron-hole. Anincrease in the number of excitons could simultaneously increase the electron/electron-hole recombination, thus the PL intensity.

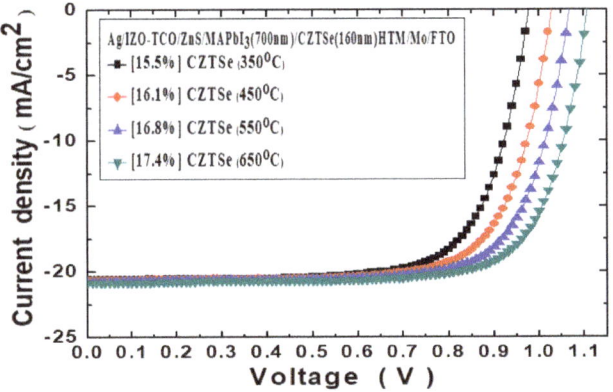

Figure 10. The J–V curves of Ag/IZO/ZnS/MAPbI$_3$/Cu$_2$ZnSnSe$_4$/Mo/FTO nanostructured solar cells, under 100 mW/cm^2 illumination. The IZO layer thickness has been 100 nm.

Table 3. The PV characteristic parameters of the Ag/IZO/ZnS/MAPbI$_3$/Cu$_2$ZnSnSe$_4$/Mo/FTO nanostructured solar cells.

Cu$_2$ZnSnSe$_4$ HTM Annealing Temperature (°C)	V_{OC} (V)	J_{SC} (mA/cm^2)	FF (%)	Eff (%)	R_s (Ω)	P_{max} (mW)
350	0.97	20.5	77.6	15.5	20.2	1.55
450	1.02	20.6	76.4	16.1	19.7	1.61
550	1.06	20.7	76.3	16.8	18.2	1.68
650	1.10	20.8	76.3	17.4	17.1	1.74

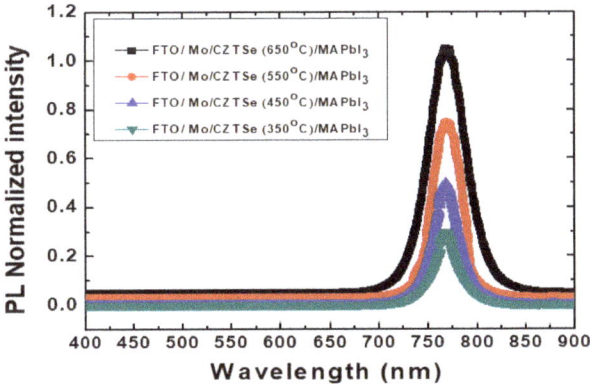

Figure 11. PL spectra of MAPbI$_3$ perovskite nano-films on Cu$_2$ZnSnSe$_4$/Mo/FTO following the various thermal annealing temperatures.

Figure 12 displays the external quantum efficiency (EQE) spectrum measurement results based on Ag/IZO/ZnS/MAPbI$_3$/Cu$_2$ZnSnSe$_4$/Mo/FTO nanostructured solar cells after the various annealing temperatures. The EQE curves for the planar solar cells have been well increased in the wavelength range of 300~800 nm from the 350 °C to 650 °C samples. Among the results of these measurements, the maximum EQE value reached nearly 85% at 550 nm for the 650 °C sample, as compared to 63% for the 350 °C sample. The higher EQE value could suggest a reduction of recombination centers. Planar solar cells involved translational electrical field distribution. It was determined entirely by 1-dimensional resonance.

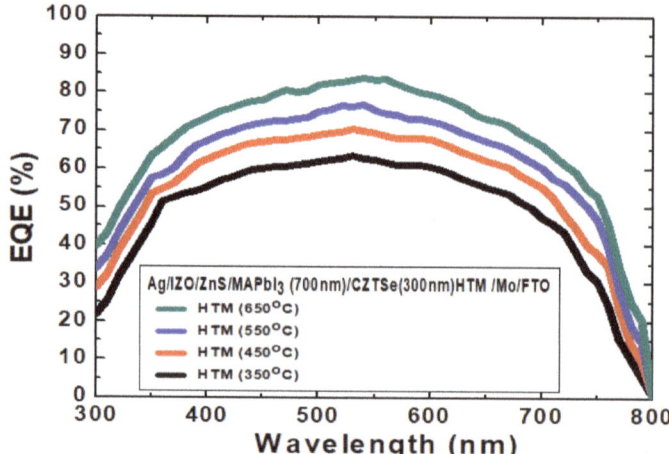

Figure 12. The EQE spectra based on Ag/IZO/ZnS/MAPbI$_3$/Cu$_2$ZnSnSe$_4$/Mo/FTO nanostructured solar cells.

Figure 13 shows the cross-sectional SEM micrograph of the Ag/IZO/ZnS/MAPbI$_3$/Cu$_2$ZnSnSe$_4$/Mo/FTO nanostructures on glass substrate. This sample has been annealed at 650 °C. The MAPbI$_3$ perovskite solar cell was evidenced with a glossy cross-section, contained few pinholes and micro-crack type of defects. The planar solar cell had uniform layer distribution through the magnetron sputtering technique and the spin-coating process. The thickness of each constituent layer could be clearly identified. Consequently, the nano-crystals were of high quality from this fabrication process, and the PV cell characteristics should be warranted in this study. In addition, Figure 14 shows the SEM micrographs of MAPbI$_3$ perovskite films on Cu$_2$ZnSnSe$_4$/Mo/FTO following the various thermal annealing temperatures. The crystal grains exhibited good surface coverage and were significantly increased in size with the increased annealing temperature.

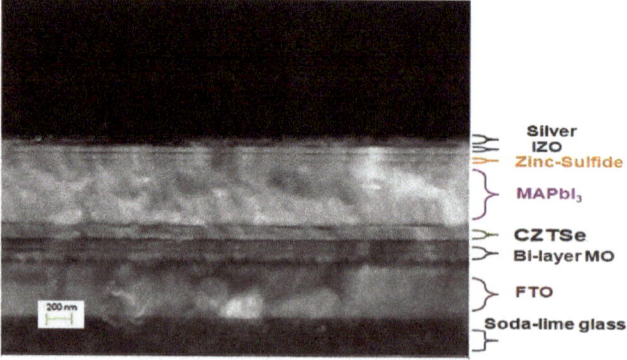

Figure 13. SEM cross-sectional micrograph of the Ag/IZO/ZnS/MAPbI$_3$/Cu$_2$ZnSnSe$_4$/Mo/FTO nanostructures on glass substrate.

Figure 15 demonstrates the secondary ion mass spectrometry (SIMS) depth profile of the Cu$_2$ZnSnSe$_4$ HTM nano-film grown on bi-layer Mo/FTO glass substrate, using ESCA PHI-5000 system (Ulvac-PHI, Kanagawa, Japan). This sample has been annealed at 650 °C. The addition of Cu-based Cu$_2$ZnSnSe$_4$ material, including selenide and MoSe$_2$ inter-film, had beneficial effect on the power conversion efficiency. The SIMS analysis provided quantitative depth profiling with good depth resolution. These features became essential to characterize the ultra-thin $Cu_2ZnSnSe_4$ HTM solar cells, where possible variations on structure or composition could lead to significant changes. In this study, $Cu_2ZnSnSe_4$ HTM film was prepared by magnetron sputtering with 160 nm in thickness, and bi-layer Mo back contact layer at about 200 nm on FTO glass substrate, all well evidenced in the depth profile. It has been noted that the use of a large collection area would provide a more representative sampling of the results. Nevertheless, the SIMS depth profile can be reconstructed after the deposition analysis to provide compositional changes at different locations.

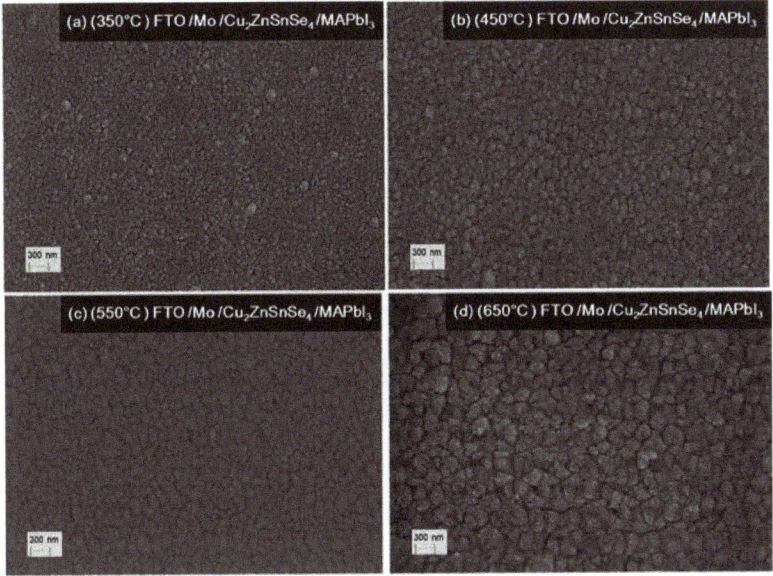

Figure 14. SEM micrographs of $MAPbI_3$ perovskite films on $Cu_2ZnSnSe_4$/Mo/FTO following the various thermal annealing temperatures: (**a**) 350 °C; (**b**) 450 °C; (**c**) 550 °C; (**d**) 650 °C.

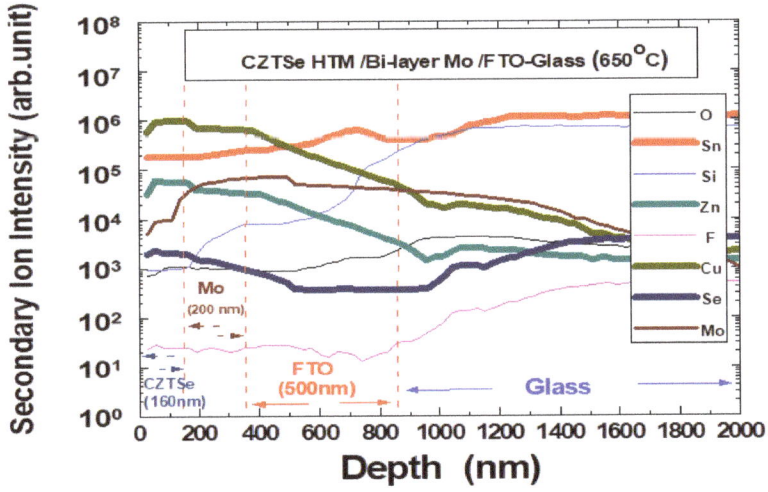

Figure 15. SIMS depth profile of the $Cu_2ZnSnSe_4$ HTM nano-film grown on bi-layer Mo/FTO glass.

4. Conclusions

In summary, one-step magnetron sputtered $Cu_2ZnSnSe_4$ nano-films have been successfully applied as novel Cu-based inorganic HTM for $MAPbI_3$ perovskite nanostructured photovoltaics. The adequate control in nano-film quality significantly improved the PV characteristic parameters of $Ag/ZnS/MAPbI_3/Cu_2ZnSnSe_4/Mo/FTO$ nanostructured solar cells. When the $Cu_2ZnSnSe_4$ HTM thickness was designed at 160 nm, and the thermal annealing temperature wasat 650 °C, its open-circuit voltage was increased to 0.98 V, and short-circuit current was increased to 20.4 mA/cm^2. The device fill factor value was increased to 71.3%, the power-conversion efficiency value was elevated to 14.3%, and the output power P_{max} value was enhanced to 1.43 mW. Furthermore, the additional inclusion of transparent conductive IZO could further enhance the PV characteristic parameters of the $Ag/IZO/ZnS/MAPbI_3/Cu_2ZnSnSe_4/Mo/FTO$ nanostructured solar cells. The open-circuit voltage was enhanced to 1.10 V, and the short-circuit current was increased to 20.8 mA/cm^2. The device fill factor value was improved to 76.3%, the power-conversion efficiency value was increased to 17.4%, the device series-resistance was decreased to 17.1 Ω., and the device output power P_{max} value could be enhanced to 1.74 mW. Therefore, the $Cu_2ZnSnSe_4$ HTM used in this study would help the development of perovskite PV technology.

Author Contributions: Methodology, C.-C.T.; investigation, G.W. and M.-J.J.; resources, D.W.C., L.-C.C. and K.-L.L.; writing—original draft preparation, C.-C.T.; writing—review and editing, G.W.; supervision, L.-B.C. and W.-S.F. All authors have read and agreed to the published version of the manuscript.

Funding: This research was funded in part by the Ministry of Science and Technology under research grants MOST105-2221-E182-059-MY3 and MOST108-2221-E182-020-MY3. The APC was funded by CGMH financial support of CMRPD3G0063 and CMRP246.

Conflicts of Interest: The authors declare no conflict of interest.

References

1. Safari, Z.; Zarandi, M.B.; Giuri, A.; Bisconti, F.; Carallo, S.; Listorti, A.; Corcione, C.E.; Nateghi, M.R.; Rizzo, A.; Colella, S. Optimizing the interface between hole transporting material and nanocomposite for highly efficient perovskite solar cells. *Nanomaterials* **2019**, *9*, 1627. [CrossRef] [PubMed]
2. Jeon, N.J.; Noh, J.H.; Yang, W.S.; Kim, Y.C.; Ryu, S.; Seo, J.; Seok, S.I. Compositional engineering of perovskite materials for high-performance solar cells. *Nature* **2015**, *517*, 476–480. [CrossRef] [PubMed]
3. Wang, T.; Zhang, H.; Hou, S.; Zhang, Y.; Li, Q.; Zhang, Z.; Gao, H.; Mao, Y. Facile synthesis of methylammonium lead iodide perovskite with controllable morphologies with enhanced luminescence performance. *Nanomaterials* **2019**, *9*, 1660. [CrossRef] [PubMed]
4. Masood, M.T.; Qudsia, S.; Hadadian, M.; Weinberger, C.; Nyman, M.; Ahlang, C.; Dahlstrom, S.; Liu, M.; Vivo, P.; Osterbacka, R.; et al. Investigation of well-defined pinholes in TiO_2 electron selective layers used in planar heterojunction perovskites solar cells. *Nanomaterials* **2020**, *10*, 181. [CrossRef] [PubMed]
5. Bresolin, B.M.; Hammouda, S.B.; Sillanpaa, M. An emerging visible-light organic-inorganic hybrid perovskite for photocatalytic applications. *Nanomaterials* **2020**, *10*, 115. [CrossRef] [PubMed]
6. Stranks, S.D.; Eperon, G.E.; Grancini, G.; Menelaou, C.; Alcocer, M.J.; Leijtens, T.; Herz, L.M.; Petrozza, A.; Snaith, H.J. Electron-hole diffusion lengths exceeding 1 micrometer in an organometal trihalide perovskite absorber. *Science* **2013**, *342*, 341–344. [CrossRef] [PubMed]
7. Kim, H.S.; Lee, C.R.; Im, J.H.; Lee, K.B.; Moehl, T.; Marchioro, A.; Moon, S.J.; Humphry-Baker, R.; Yum, J.H.; Moser, J.E.; et al. Lead iodide perovskite sensitized all-solid-state submicron thin film mesoscopic solar cell with efficiency exceeding 9%. *Sci. Rep.* **2012**, *2*, 591. [CrossRef]
8. Jeng, J.Y.; Chiang, Y.F.; Lee, M.H.; Peng, S.R.; Guo, T.F.; Chen, P.; Wen, T.C. $CH_3NH_3PbI_3$ perovskite/fullerene planar-heterojunction hybrid solar cells. *Adv. Mater.* **2013**, *25*, 3727–3732. [CrossRef]
9. Choi, H.; Mai, C.K.; Kim, H.B.; Jeong, J.; Song, S.; Bazan, G.C.; Kim, J.Y.; Heeger, A.J. Conjugated polyelectrolyte hole transport layer for inverted-type perovskite solar cells. *Nat. Commun.* **2015**, *6*, 7348. [CrossRef]
10. Battaglia, C.; Cuevas, A.; De Wolf, S. High-efficiency crystalline silicon solar cells: Status and perspectives. *Energy Environ. Sci.* **2016**, *9*, 1552–1576. [CrossRef]

11. Wu, Q.; Xue, C.; Li, Y.; Zhou, P.; Liu, W.; Zhu, J.; Dai, S.; Zhu, C.; Yang, S. Kesterite Cu$_2$ZnSnS$_4$ as a low-cost inorganic hole-transporting material for high-efficiency perovskite solar cells. *ACS Appl. Mater. Interfaces* **2015**, *7*, 28466–28473. [CrossRef] [PubMed]
12. Ramanujam, J.; Singh, U.P. Copper indium gallium selenide based solar cells—A review. *Energy Environ. Sci.* **2017**, *10*, 1306–1319. [CrossRef]
13. Van Lare, C.; Yin, G.; Polman, A.; Schmid, M. Light coupling and trapping in ultrathin Cu(In,Ga)Se$_2$ solar cells using dielectric scattering patterns. *ACS Nano* **2015**, *9*, 9603–9613. [CrossRef] [PubMed]
14. Li, Z.; Klein, T.R.; Kim, D.H.; Yang, M.; Berry, J.J.; Van Hest, M.; Zhu, K. Scalable fabrication of perovskite solar cells. *Nat. Rev. Mater.* **2018**, *3*, 18017. [CrossRef]
15. Christians, J.A.; Schulz, P.; Tinkham, J.S.; Schloemer, T.H.; Harvey, S.P.; De Villers, B.J.; Sellinger, A.; Berry, J.J.; Luther, J.M. Tailored interfaces of unencapsulated perovskite solar cells for >1000 h operational stability. *Nat. Energy* **2017**, *3*, 68–74. [CrossRef]
16. Jackson, P.; Wuerz, R.; Hariskos, D.; Lotter, E.; Witte, W.; Powalla, M. Effects of heavy alkali elements in Cu(In,Ga)Se$_2$ solar cells with efficiencies up to 22.6%. *Phys. Status Solidi RRL* **2016**, *10*, 583–586. [CrossRef]
17. Wen, X.; Chen, C.; Lu, S.; Li, K.; Kondrotas, R.; Zhao, Y.; Chen, W.; Gao, L.; Wang, C.; Zhang, J.; et al. Vapor transport deposition of antimony selenide thin film solar cells with 7.6% efficiency. *Nat. Commun.* **2018**, *9*, 2179. [CrossRef]
18. Sapkota, Y.R.; Mazumdar, D. Bulk transport properties of bismuth selenide thin films grown by magnetron sputtering approaching the two-dimensional limit. *J. Appl. Phys.* **2018**, *124*, 105306. [CrossRef]
19. Ong, K.H.; Agileswari, R.; Maniscalco, B.; Arnou, P.; Kumar, C.C.; Bowers, J.W.; Marsadek, M.B. Review on substrate and molybdenum back contact in CIGS thin film solar cell. *Int. J. Photoenergy* **2018**, *2018*, 9106269. [CrossRef]
20. Khoshsirat, N.; Ali, F.; Tiong, V.T.; Amjadipour, M.; Wang, H.; Shafiei, M.; Motta, N. Optimization of Mo/Cr bilayer back contacts for thin-film solar cells. *Beilstein J. Nanotechnol.* **2018**, *9*, 2700–2707. [CrossRef]
21. Dalapati, G.K.; Zhuk, S.; Masudy-Panah, S.; Kushwaha, A.; Seng, H.L.; Chellappan, V.; Suresh, V.; Su, Z.; Batabyal, S.K.; Tan, C.C.; et al. Impact of molybdenum out diffusion and interface quality on the performance of sputter grown CZTS based solar cells. *Sci. Rep.* **2017**, *7*, 1350. [CrossRef] [PubMed]
22. Borri, C.; Calisi, N.; Galvanetto, E.; Falsini, N.; Biccari, F.; Vinattieri, A.; Cucinotta, G.; Caporali, S. First proof-of-principle of inorganic lead halide perovskites deposition by magnetron-sputtering. *Nanomaterials* **2020**, *10*, 60. [CrossRef] [PubMed]
23. Yin, L.; Cheng, G.; Feng, Y.; Li, Z.; Yang, C.; Xiao, X. Limitation factors for the performance of kesterite Cu$_2$ZnSnS$_4$ thin film solar cells studied by defect characterization. *RSC Adv.* **2015**, *5*, 40369–40374. [CrossRef]
24. Chalapathy, R.B.V.; Jung, G.S.; Ahn, B.T. Fabrication of Cu$_2$ZnSnS$_4$ films by sulfurization of Cu/ZnSn/Cu precursor layers in sulfur atmosphere for solar cells. *Sol. Energy Mater. Sol. Cells* **2011**, *95*, 3216–3221. [CrossRef]
25. Yin, G.; Knight, M.W.; Van Lare, M.C.; Garcia, M.M.S.; Polman, A.; Schmid, M. Optoelectronic enhancement of ultrathin CuIn$_{1-x}$Ga$_x$Se$_2$ solar cells by nanophotonic contacts. *Adv. Opt. Mater.* **2017**, *5*, 1600637. [CrossRef]
26. Khojier, K.; Mehr, M.R.; Savaloni, H. Annealing temperature effect on the mechanical and tribological properties of molybdenum nitride thin films. *J. Nanostruct. Chem.* **2013**, *3*, 5. [CrossRef]
27. Aqil, M.M.; Azam, M.A.; Aziz, M.F.; Latif, R. Deposition and characterization of molybdenum thin film using direct current magnetron and atomic force microscopy. *J. Nanotechnol.* **2017**, *2017*, 4862087. [CrossRef]
28. Chen, C.; Zhang, L.; Shi, T.; Liao, G.; Tang, Z. Controllable synthesis of all inorganic lead halide perovskite nanocrystals with various appearances in multiligand reaction system. *Nanomaterials* **2019**, *9*, 1751. [CrossRef]
29. Huang, P.C.; Sung, C.C.; Chen, J.H.; Huang, C.H.; Hsu, C.Y. The optimization of a Mo bi-layer and its application in Cu(In,Ga)Se$_2$ solar cells. *Appl. Surf. Sci.* **2017**, *425*, 24–31. [CrossRef]
30. Shin, B.; Zhu, Y.; Bojarczuk, N.A.; Chey, S.J.; Guha, S. Control of an interfacial MoSe$_2$ layer in Cu$_2$ZnSnSe$_4$ thin film solar cells: 8.9% power conversion efficiency with a TiN diffusion barrier. *Appl. Phys. Lett.* **2012**, *101*, 053903. [CrossRef]
31. Lin, Y.C.; Liu, K.T.; Hsu, H.R. A comparative investigation of secondary phases and MoSe$_2$ in Cu$_2$ZnSnSe$_4$ solar cells: Effect of Zn/Sn ratio. *J. Alloy. Compd.* **2018**, *743*, 249–257. [CrossRef]
32. Nishimura, T.; Sugiura, H.; Nakada, K.; Yamada, A. Characterization of interface between accurately controlled Cu-deficient layer and Cu(In,Ga)Se$_2$ absorber for Cu(In,Ga)Se$_2$ solar cells. *Phys. Status Solidi RRL* **2018**, *12*, 1800129. [CrossRef]

33. Zhang, X.; Kobayashi, M.; Yamada, A. Comparison of Ag(In,Ga)Se$_2$/Mo and Cu(In,Ga)Se$_2$/Mo interfaces in solar cells. *ACS Appl. Mater. Interfaces* **2017**, *9*, 16215–16220. [CrossRef] [PubMed]
34. Zheng, E.; Wang, Y.; Song, J.; Wang, X.F.; Tian, W.; Chen, G.; Miyasaka, T. ZnO/ZnS core-shell composites for low-temperature-processed perovskite solar cells. *J. Energy Chem.* **2018**, *27*, 1461–1467. [CrossRef]
35. Ke, W.; Stoumpos, C.C.; Logsdon, J.L.; Wasielewski, M.R.; Yan, Y.; Fang, G.; Kanatzidis, M.G. TiO$_2$-ZnS cascade electron transport layer for efficient formamidinium tin iodide perovskite solar cells. *J. Am. Chem. Soc.* **2016**, *138*, 14998–15003. [CrossRef] [PubMed]
36. Liu, J.; Gao, C.; Luo, L.; Ye, Q.; He, X.; Ouyang, L.; Guo, X.; Zhuang, D.; Liao, C.; Mei, J.; et al. Low-temperature, solution processed metal sulfide as an electron transport layer for efficient planar perovskite solar cells. *J. Mater. Chem. A* **2015**, *3*, 11750–11755. [CrossRef]
37. Yuan, H.; Zhao, Y.; Duan, J.; Wang, Y.; Yang, X.; Tang, Q. All-inorganic CsPbBr3 perovskite solar cell with 10.26% efficiency by spectra engineering. *J. Mater. Chem. A* **2018**, *6*, 24324–24329. [CrossRef]
38. Shyju, T.S.; Anandhi, S.; Suriakarthick, R.; Gopalakrishnan, R.; Kuppusami, P. Mechanosynthesis, deposition and characterization of CZTS and CZTSe materials for solar cell applications. *J. Solid State Chem.* **2015**, *227*, 165–177. [CrossRef]

© 2020 by the authors. Licensee MDPI, Basel, Switzerland. This article is an open access article distributed under the terms and conditions of the Creative Commons Attribution (CC BY) license (http://creativecommons.org/licenses/by/4.0/).

Article

Influence of Thickness and Sputtering Pressure on Electrical Resistivity and Elastic Wave Propagation in Oriented Columnar Tungsten Thin Films

Asma Chargui [1], Raya El Beainou [1], Alexis Mosset [1], Sébastien Euphrasie [1], Valérie Potin [2], Pascal Vairac [1] and Nicolas Martin [1,*]

[1] Institut FEMTO-ST, UMR 6174, CNRS, ENSMM, Univ. Bourgogne Franche-Comté, 15B, Avenue des Montboucons, 25030 BESANCON CEDEX, France; asma.chargui@femto-st.fr (A.C.); raya.elbeainou@femto-st.fr (R.E.B.); alexis.mosset@femto-st.fr (A.M.); sebastien.euphrasie@femto-st.fr (S.E.); pascal.vairac@femto-st.fr (P.V.)

[2] Laboratoire Interdisciplinaire Carnot de Bourgogne (ICB), UMR 6303, CNRS, Univ. Bourgogne Franche-Comté, 9, Avenue Alain Savary, BP 47 870, F-21078 DIJON CEDEX, France; Valerie.Potin@u-bourgogne.fr

* Correspondence: nicolas.martin@femto-st.fr; Tel.: +33-3-6308-2431

Received: 9 December 2019; Accepted: 25 December 2019; Published: 1 January 2020

Abstract: Tungsten films were prepared by DC magnetron sputtering using glancing angle deposition with a constant deposition angle $\alpha = 80°$. A first series of films was obtained at a constant pressure of 4.0×10^{-3} mbar with the films' thickness increasing from 50 to 1000 nm. A second series was produced with a constant thickness of 400 nm, whereas the pressure was gradually changed from 2.5×10^{-3} to 15×10^{-3} mbar. The A15 β phase exhibiting a poor crystallinity was favored at high pressure and for the thinner films, whereas the bcc α phase prevailed at low pressure and for the thicker ones. The tilt angle of the columnar microstructure and fanning of their cross-section were tuned as a function of the pressure and film thickness. Electrical resistivity and surface elastic wave velocity exhibited the highest anisotropic behaviors for the thickest films and the lowest pressure. These asymmetric electrical and elastic properties were directly connected to the anisotropic structural characteristics of tungsten films. They became particularly significant for thicknesses higher than 450 nm and when sputtered particles were mainly ballistic (low pressures). Electronic transport properties, as well as elastic wave propagation, are discussed considering the porous architecture changes vs. film thickness and pressure.

Keywords: sputtering; GLAD; tilted columns; anisotropy; electrical resistivity; elastic wave propagation

1. Introduction

Among the solid-state physical properties, understanding the propagation of electrons and elastic waves in materials remains a challenging task. It was commonly pointed out that the control of transport mechanisms in solids greatly influences resulting applications in devices used as bulk and surface acoustic wave systems (BAW and SAW), semiconductors, gas sensors, and so on [1–3]. It is also admitted that the crystal symmetries mainly affect the direction-dependent properties of materials due to an intrinsic anisotropy at the atomic scale coming from spatial differences between bonds in the crystal structure. Such anisotropy becomes negligible or even vanishes when the material becomes polycrystalline, which is particularly true when the material dimension changes from 3D to 2D. As a result of the loss of this intrinsic anisotropy, structuring at the micro- and nanoscale appears as an attractive strategy to produce directional behaviors in surfaces and thin solid films [4–7].

These spatially organized surfaces and films may exhibit asymmetric characteristics in electronic and ionic transports, electromagnetic wave propagations, magnetic properties, or even mechanical and tribological performances [8–10].

For the last decades, tuning the physical properties of thin films and surfaces and producing anisotropic behaviors have attracted many researchers since it allows producing multifunctionality within the same material [11,12]. Then, the structuring of thin solid films at the micro and nanoscale using top-down or bottom-up strategies has become a pertinent tool to get original patterns and designs favoring anisotropic behaviors. Among these strategies, growth of thin films by vacuum processes, such as magnetron sputtering by the GLAD technique (GLancing Angle Deposition), recently became a very motivating way to create original architectures (tilted columns, zigzags, spirals, etc. [13–16]) and thus, to develop anisotropic properties [17–19]. However, some scientific and technological challenges are still to be addressed, such as the tunability of anisotropic characteristics and understanding the correlations between created structures and propagation of waves, electrons, ions, or temperature through these structured thin films. In addition, growth conditions by sputtering (pressure, temperature, gases, etc.) may strongly influence some characteristics and the final structure of as-deposited thin films. These experimental parameters restrain some achievable properties of thin films, and they still need to be explored, particularly in the GLAD technique where the growth mechanisms may largely differ compared to conventional sputtering [20,21].

In this article, we report on tilted columnar tungsten thin films sputter-deposited by magnetron sputtering using the GLAD technique with a constant deposition angle of $\alpha = 80°$. Two deposition parameters have been studied: the film's thickness and the argon sputtering pressure giving rise to two series of films, i.e., a first series with constant pressure and a growing film thickness, and a second series with a constant thickness and different pressures. Electrical resistivity and surface elastic wave propagation have been systematically investigated for both series. This study is motivated by the understanding of anisotropic behaviors in terms of electronic transport properties and elastic wave propagation in nanostructured thin films exhibiting a tilted columnar architecture. The choice of tungsten films is due to its ability to produce a columnar structure with tunable cross-section morphologies depending on deposition time and sputtering conditions. By increasing the film's thickness, it has been shown that the electrical resistivity and elastic wave velocities were both reduced, whereas anisotropy was favored. A higher pressure produced more resistive films with an increase in the elastic wave velocity, where anisotropic behaviors were reduced. The evolution of these physical properties as a function of the film's thickness and argon sputtering pressure is discussed, taking into account the evolution of the crystallographic structure and the films' morphology at the micro and nanoscale.

2. Materials and Methods

Tungsten films were sputter-deposited on glass and (100) Si substrate by DC magnetron sputtering from a pure metallic target (51 mm diameter and 99.9 at.% purity) in a home-made vacuum chamber. The experimental deposition system was a 40 L sputtering chamber pumped down via a turbo-molecular pump backed by a primary pump leading to a residual vacuum of 10^{-8} mbar. The target current was fixed at 100 mA, and the target-to-substrate distance was 65 mm. No external heating was applied during the growth stage, and depositions were carried out at room temperature. The GLAD (GLancing Angle Deposition) technique [22] was implemented to produce tilted columnar architectures. It consists of depositing thin films under conditions of obliquely incident flux of the sputtered particles and on a fixed or mobile (rotating) substrate. In this study, the substrate is tilted at an angle $\alpha = 80°$ without rotating (fixed substrate). Two series of samples were produced. For the first series, an argon flow rate of 5.6 sccm and a constant pumping speed of 24 L s^{-1} were used. These conditions produced an argon sputtering pressure of 4.0×10^{-3} mbar, and the deposition time was systematically changed to get a film thickness variation from 50 to 1000 nm. For the second series, the deposition time was set to deposit around 400 nm thick films changing the argon sputtering pressure from 2.5×10^{-3} up to

15×10^{-3} mbar (pumping speed and argon flow rate were both adjusted to get the required range of pressures).

The crystallographic structure of W films was characterized by X-ray diffraction (XRD). Measurements were carried out using a Bruker D8 focus diffractometer with a Cobalt X-ray tube (Co $\lambda_{K\alpha 1} = 0.178897$ nm) in a $\theta/2\theta$ configuration. Patterns were recorded with a step of 0.02° per 0.2 s and a 2θ angle ranging from 20° to 80°. Scanning electron microscopy (SEM) was used to view the surface and the fractured cross-section of the films with a JEOL JSM 7800 field emission SEM. DC electrical resistivity of the films was measured at room temperature in air by the four-probe van der Pauw method. The measurements were carried out on glass substrates, and the surface anisotropic electrical resistivity was determined using the method previously developed by Bierwagen et al. [23].

A femtosecond pump-probe setup was used for the measurement of the elastic wave propagation. This technique is based on an ultrashort laser pump pulse that interacts with the surface of the sample. This leads to a sudden temperature rise that will excite the surface into vibration through the thermoelastic effect. The subsequent reflectivity modifications are measured by a low-power second laser probe pulse, with an energy around a tenth of the pump pulse energy. Two femtosecond Ytterbium lasers (T-Pulse Duo from Amplitude System were used in our setup with a heterodyne configuration also called asynchronous optical sampling technique (ASOPS), where an increasing delay between the pump and probe pulses was induced by a small frequency shift of their repetition rates [24]. This frequency shift was 700 Hz with a repetition rate of 48 MHz. This gave rise to an additional time difference between the pump and the probe beam of around 300 ps every pulse. The whole time spanning 21 ns was thus obtained with a resolution of about 1 ps. The average pump and probe beam powers were 5 and 0.5 mW, respectively. Both beams were focused on the sample with a 1 µm spot radius. Thanks to a lens mounted on a 2D translation stage in the pump optical path, the pump-probe distance could be scanned to image the reflectivity. From this imaging, we deduced the dispersion curves obtained by computing the 2D-FFT (2 Dimensions Fast Fourier Transform) of the relative reflectivity vs. pump-probe distance and time. The elastic wave group velocities were obtained from the local slope of the dispersion curves (frequency vs. wave number/2π). They were calculated at the most intense normalized reflectivity of the pseudo-Rayleigh mode, which was at the wave number $k = 2\pi \times 3 \times 10^5$ m^{-1} [24].

We used a 3D finite element (FE) model to correlate the porosity with the measured dispersion curves; finite element analyses were performed using COMSOL Multiphysics. The simulation model was based on a silicon block on which lay a nanoporous tungsten film with periodic parallelepipedic holes to model the macroporosity (visible holes in the SEM pictures). The structure was simulated to be infinitely periodic with Bloch–Floquet conditions, both in x and y-directions. This model is very simplistic but gives the overall tendencies and is a computer-based tool, which does not involve a strong memory requirement.

Figure 1 shows the unit cell assumed in this paper.

It consisted of a square cuboid silicon block (parallelepiped with two opposite square faces) with a side $a = 500$ nm and height $h_{Si} = 25$ µm (not on scale in Figure 1), on which lay a porous tungsten film. A hole with a rectangular base (length c and width d) was dug in the tungsten film. The side a (period) was chosen from the SEM pictures. A previous study [24] showed that the anisotropy is connected to the geometry of the holes, especially their form factor and the choice of the period, the column tilt angle having only a limited influence. The material constants used in the simulations were from literature data [25] for the silicon and modified for the tungsten to take into account the nanoporosity.

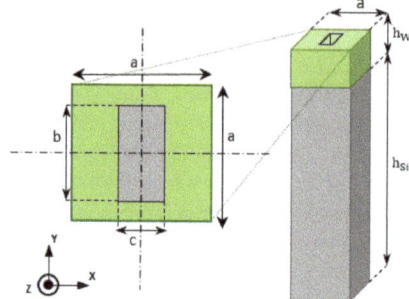

Figure 1. Scheme of the square cuboid unit cell (lattice constant *a*) composed of Si substrate (grey; thickness h_{Si}) and tungsten film (green; thickness h_W). Position and dimensions (length *c* and width *d*) of the parallelepipedic hole dug in the tungsten film are indicated.

Since Biot's theory [26] for porous solids, several groups have studied the porosity as a key parameter to establish a theory about elastic wave propagation in a system composed of a porous elastic solid. Investigations focused on doped porous Si wafers [27,28] propose an empirical relation between the longitudinal and shear waves (bulk waves) and the porosity following:

$$v = v_0(1-p)^k, \qquad (1)$$

where v is the velocity of the wave in the porous material (m s^{-1}), v_0 the velocity of the bulk wave in the bulk material (m s^{-1}), p the porosity of the material and k a parameter, close to one, depending on the nature of the wave, the porosity, and the material. This form of velocity dependence on porosity was used to fit our experimental results with $k = 1$. Neglecting the density of air compared to the density of as-deposited film, the density ρ (kg m^{-3}) of the nanoporous film is given by:

$$\rho = (1-p) \times \rho_0, \qquad (2)$$

where ρ_0 (kg m^{-3}) is the density of bulk material. Thus, the elastic constants vary with nearly the 3rd power of $(1-p)$ [29].

The lattice constant *a* of the unit cell was chosen from the SEM observations, the size of the hole (*b* and *c* parameters), and the nanoporosity p were adjusted to fit the experimental data. The global porosity π (including the nanoporosity p and the macroporosity) is given by:

$$\pi = p + (1-p)bc/a^2 \qquad (3)$$

3. Results and Discussion

3.1. First Series: Thickness from 50 to 1000 nm

3.1.1. Morphology and Structure

SEM observations of as-deposited W films exhibited a gradual change in surface and cross-section morphologies as a function of the film's thickness (Figure 2). For the thinnest film (50 nm), no clear shapes could be distinguished from the top view, whereas very small columns (about a few tens nm width) could be seen from the cross-section view (Figure 2a). Despite the very low film thickness, they were oriented in the direction of the incoming particle flux. For this first series, the sputtering pressure was 4.0×10^{-3} mbar, which mostly produced ballistic sputtered particles. The incident vapor was thus highly directional, and the shadowing effect became effective from the early growing stage, i.e., after a thickness of a few nanometers. Randomly distributed islands were observed when the thickness was over 100 nm, and tilted columns (β close to 39° ± 2°) were even more distinct (Figure 2b).

A further increase in thickness from 200 to 1000 nm led to a better-defined microstructure (Figure 2c–f). Top views showed a more and more corrugated surface as the thickness increases. Asymmetric and elongated voids alternated with columns exhibiting an elliptical cross-section following the direction perpendicular to the particle flux, i.e., y-direction. This anisotropic microstructure has ever been reported for tungsten and other metallic thin films prepared by GLAD [30]. This column fanning was closely connected to the film growth. During the first growing stages, nuclei were randomly distributed on the substrate surface with no clear structure. As the deposition progresses, the column features gradually fanned out along the y-axis (normal to the particle flux) and the shadowing phenomenon started acting. As a result, a transverse growth was favored leading to a well-defined elliptical shape of the columns section and a rising column interspacing with a more voided structure.

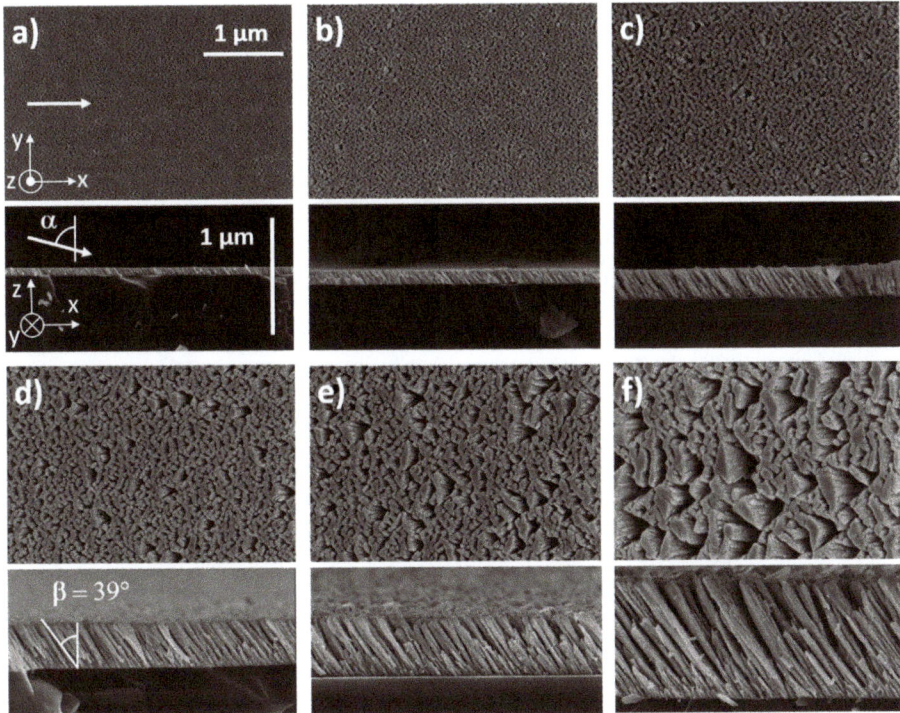

Figure 2. Top and cross-section views performed by SEM of W films sputter-deposited with a substrate angle $\alpha = 80°$ and an argon sputtering pressure $P_{Ar} = 4.0 \times 10^{-3}$ mbar. White arrows indicate the direction of incoming particle flux. Column angle β was nearly constant (39° ± 2°) whatever the film's thickness, which was systematically increased from: (a) 50; (b) 100; (c) 200; (d) 450; (e) 600, and (f) 1000 nm.

Cross-section observations of slanted columnar W structures also showed a continuous evolution as a function of the film's thickness. Some columns fell under the shadow of bigger ones and become extinct (especially noticeable from the cross-section views of the thickest films in Figure 2e,f). This shadowing-induced competition between columns continuously occurred during the film growth. It was previously shown that the GLAD films morphology is scale-invariant with a power-law scaling connecting thickness and columns' width in x- and y-directions [31]. Surface self-diffusion, atomic mobility, ballistic character of incoming particles, or dragging phenomenon are quite a few mechanisms which significantly influence the column broadening and tilting vs. film thickness, and thus, the final structural morphology and anisotropy [32]. Despite several growth simulations in agreement with experimental data and predicting column angle and broadening, the structural anisotropy of GLAD films still remains strongly dependent on operating conditions and experimental methods (cf. Section 3.2 on the role of the argon sputtering pressure).

XRD measurements also showed that the crystalline structure of W GLAD films was meaningfully affected by the film's thickness, as reported in Figure 3.

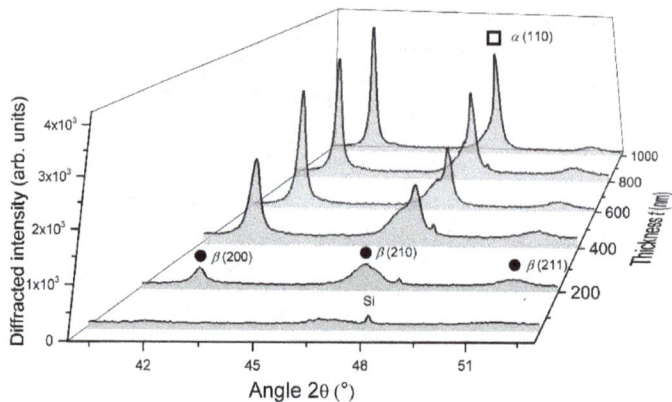

Figure 3. XRD patterns of W thin films sputter-deposited on (100) Si substrate with an angle $\alpha = 80°$, an argon sputtering pressure of 4.0×10^{-3} mbar and for a thickness changing from 50 to 1000 nm. Only the 41°–52° range is presented since no significant peaks were measured elsewhere. Diffracted signals showed the occurrence of α and β phases (π and λ, respectively).

For the smallest thickness (50 nm), no significant diffracted signals were recorded but only a broad and poorly intense band located at 2θ angle close to 47°, which is related to the (210) planes of the metastable A15 β phase. As the film's thickness reached 200 nm, more intense peaks were clearly measured and are once again assigned to the β phase. Peaks became even more intense and narrow for thicknesses higher than 450 nm. The bcc α phase appeared with an important diffracted signal corresponding to (110) planes recorded at $2\theta = 47.12°$. Thus, increasing the film's thickness favors the crystallinity of α and β phases in the columnar structure. From α (110) and β (200) peaks, the crystal size determined using the Scherrer formula reached 31 and 57 nm for α and β phases, respectively, as the thickness reached 1000 nm. In addition to the α phase occurrence, the β phase grew with a preferential orientation along (200) direction. As a result, a growth competition occurred between these two W phases as a function of the film's thickness. These results can be connected to former investigations focused on the crystallographic structure of W films prepared by conventional sputtering (i.e., deposition angle $\alpha = 0°$) [33]. During the early film deposition stage, β phase nuclei are initially formed up to a so-called critical thickness, and a phase mixture is produced (appearance of the α phase). Such a critical thickness is in the order of a few to several tens nanometers and largely depends on the experimental growth parameters, especially pressure, deposition rate, and power [34]. Depositing above this thickness by a conventional sputtering process commonly leads to a β to α phase transformation due to a diffusion-controlled process of W atoms [35]. In the GLAD sputtering technique, growth mechanisms may differ since the direction of the W particle flux can yield dense or fibrous morphologies, as the column apexes are in front of the flux or in the shadowing zone [36]. This inhomogeneous growing evolution in the columnar growth significantly produces a completely different microstructural morphology and, thus, different crystallographic structure and properties through the cross-section of a given column. Taking into account that the β phase is favored when the energy of deposited species is reduced [37], one can expect the β phase occurrence on the column sides located in the shadowing zone, whereas the α phase prevails in front of the incoming particle flux. As a result, for thicknesses higher than 200 nm, α and β phases coexisted in W GLAD thin films sputter-deposited with our operating conditions. In addition to the β (200) peak located at 2θ = 41.58°, this phase mixture is well illustrated by the shouldered peak close to 47° indicating the presence of both phases. The latter asymmetric signal can be deconvoluted to the β (210) and α (110) Bragg peaks (2θ = 46.76° and 47.12°, respectively) to get approximately α and β phase content in the film from the selected phase peak area to the total peak area ratio [37]. For the lowest thicknesses (< 200 nm), films were mainly composed of the β phase, and the α phase started growing at 200 nm (α content was a few percentages by volume). When the film's thickness reached 1000 nm, the α phase significantly rose with a proportion by volume higher than 40%.

3.1.2. Electrical Resistivity

DC electrical resistivity of GLAD W thin films deposited on glass substrate also varied as a function of the film's thickness from 1.1×10^{-5} to 5.5×10^{-6} Ω m, as shown in Figure 4.

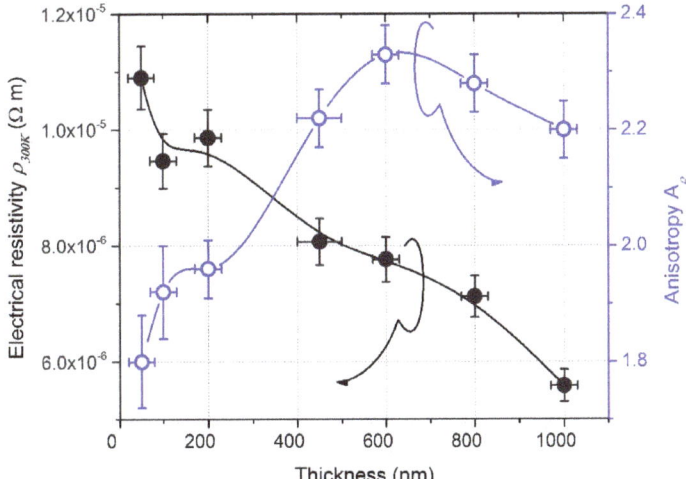

Figure 4. DC electrical resistivity and anisotropic resistivity measured at room temperature (300 K) as a function of the film's thickness.

It was higher than the bulk value ($\rho_{300K} = 5.4 \times 10^{-8}$ Ω m for bulk W [38]), which is classically reported in GLAD films [39]. This higher resistivity is assigned to the enhancement of the electron scattering by surfaces and grain boundaries induced by a much more porous structure promoted as the incident angle α tends to 90°. However, the resistivity evolution of W GLAD films did not exhibit the classical saturation effect as typically observed in conventional sputtered films when the thickness exceeds a few tens nanometers [40]. This continuous reduction of resistivity vs. thickness has to be connected to XRD results, which simultaneously showed an increase in the crystal size and favoring of the α phase as a function of the film's thickness. Since the electrical conductivity of metallic thin films is limited by the scattering of electrons at grain boundaries, an increase in the crystal size induces a longer electron mean free path and thus, improves the electrical conductivity of the films. In addition, Petroff and Reed [41] previously showed that the amounts of α and β phases present in W thin films strongly affect the resistivity because bulk α and β phases exhibit a significant difference of resistivity with $\rho_\alpha = 5.4 \times 10^{-8}$ Ω m lower than $\rho_\beta = 1.5\text{--}3.5 \times 10^{-6}$ Ω m at 300 K [42]). As a result, an increase in the crystal size associated with a larger proportion of the α phase as thickness increased, both contributed to the drop in the films' resistivity.

Figure 4 also illustrates the influence of the films' thickness on anisotropic resistivity A_ρ defined as the resistivity ratio following x and y-directions (i.e., parallel and perpendicular to the direction of the particle flux). For the lowest thickness of 50 nm, $A_\rho = 1.8$, which was surprisingly high since no clear anisotropic morphology was viewed from SEM observations in Figure 2a. On the other hand, some studies have deservedly reported that the shadowing effect may become effective from the first growing stage of oblique angle deposition [43], especially at low pressure for the sputtering process and for deposition angles α higher than 80°, which were our operating conditions. It is also worth noting that the first nuclei created nanoscale topographies, thus inducing an initial surface roughness, which is a principal requirement for the shadowing effect to begin. The latter is a key parameter, particularly when the surface diffusion of incoming particles is limited. Such is the case of W atoms where the surface self-diffusion energy $E_d = 3.10$ eV, which is high compared to other metals where E_d is lower than 1 eV [44]. Thus, atoms such as W are unable to fill-in the shadowed regions (shadowing prevents structural broadening of the columns in the direction of the particle flux). As a result, despite no clear surface structural asymmetry being clearly distinguished from SEM images of 50 nm thick films, a structural anisotropy certainly occurred shortly after the first growing stage.

Anisotropic resistivity became more and more relevant as the film's thickness increased and reached 2.3 at 600 nm. Although the film's thickness increased up to 1000 nm, A_ρ remained higher than 2.2. This anisotropic behavior, which was more obvious for thicknesses higher than a few hundred nm, agrees with other studies focused on the electronic transport properties of GLAD thin films [45]. Such behavior is mainly assigned to the elliptical shape of the columnar cross-section (fanning mechanism during the growth, as shown in Figure 2), being especially prominent in W GLAD thin films [46]. An alternation of dense and voided architecture is rather produced in the direction of the particle flux, whereas dense and chained columns were obtained following the normal direction. A further increase in the film's thickness did not enhance anisotropic resistivity. This saturation can be associated with a nearly stable crystallographic structure (α and β phase mixture) and nearly unchanged morphology (top and cross-section views by SEM rather show a scaled architecture) from 450 to 1000 nm.

3.1.3. Elastic Wave Propagation

The group velocities of the Rayleigh waves following the x- (parallel to the incoming flux) and y-direction (perpendicular to the incoming particle flux) were obtained from the local slope of the dispersion curves for a wave vector $k/2\pi = 1/\lambda = 3 \times 10^5$ m^{-1} in our case (where λ is the wavelength). The calculated group velocities of the Rayleigh waves of W thin films change as a function of the film's thickness, as shown in Figure 5.

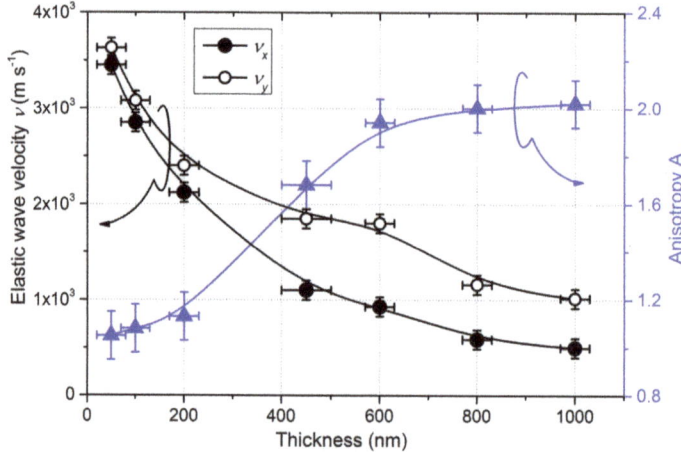

Figure 5. Group velocities of pseudo-Rayleigh waves along x and y-axes, and related anisotropic coefficient as a function of the W film's thickness.

As the film's thickness increased, velocities decreased, and the anisotropy coefficient A_v, defined as the velocity ratio following x and y-directions, increased. Since the films were smaller than the wavelength, the Rayleigh waves also propagated into the substrate. Therefore, one can expect an increase in A_v with the thickness. Moreover, as the Rayleigh velocity in silicon (4917 m s^{-1} along the <100> direction [47]) is higher than in tungsten (2646 m s^{-1} [47]), the decrease in velocities is also expected. However, analytical calculation with a homogeneous film and simulations showed that the effect of the substrate is not enough to explain these behaviors (cf. the computing of porosity at the end of this section).

For the smallest thickness (50 nm), velocities were similar along x and y-axes (3450 ± 50 m s^{-1} and 3630 ± 50 m s^{-1}, respectively). As a result, the anisotropic coefficient A_v equaled to 1.05, contrary to the resistivity anisotropy (A_ρ = 1.8). This result could be expected since a good proportion of the waves propagated into the substrate. It is worth noting that these velocities were higher than the surface wave velocities of tungsten bulk metals, a consequence of the higher velocity in Si.

Increasing the thickness up to 450 nm led to an important drop for the pseudo-Rayleigh wave velocity with v_x = 1100 m s^{-1} and v_y = 1848 m s^{-1} along x- and y-directions, respectively. This sudden decrease in velocities is mainly attributed to the formation of voided microstructure and large spaces between the inclined columns, which became more and more important as the thickness increased. Moreover, the column features gradually fanned out along the y-direction while shadowing prevented significant structural broadening along the x-direction. The cross-section thus became increasingly elliptical as deposition continued, and the basic columnar microstructure exhibited structural anisotropy because of the development of an elongated column cross-section. Therefore, the significant difference between velocities v_x and v_y is linked to the structural anisotropy, which developed in the film plane. Increasing the thickness to and above 800 nm led to a saturation of the anisotropic coefficient A_v tending to 2.0. This stabilization, as with the resistivity, comes from the stable crystallographic structure (α and β phase mixture) and the nearly unchanged and scalable morphology.

To separate the effect of the substrate from that of the film's morphology, FE simulations, including macro and nanoporosities, were performed with a simplistic model, as described in Section 2. Table 1 illustrates the evolution of the global porosity as a function of the film's thickness (fitted to the experimental data adjusting the hole sizes and porosity).

Table 1. Calculated global porosity as a function of the W film's thickness.

Thickness (± 50 nm)	100	200	450	600	800	1000
Global porosity π (± 10% of the bulk)	20	33	62	65	66	69

These results confirm that the microstructure and porosity of the film changed with the thickness. The fitted porosity steadily increased up to a thickness of 450 nm, from π = 20% for 100 nm to 62% of the bulk for 450 nm. Such high values of porosity were also obtained for thinner Si films prepared by GLAD [43]. This porous architecture produced in GLAD films is linked to the increase in the average distance between columns and the vanishing of other columns due to the shadowing effect (typical phenomenon driven by the growth competition, which is inherent to the GLAD process). For thicker films, the porosity still increased, but to a lesser extent, from π = 65% at for 600 nm to 69% of the bulk for 1000 nm. This evolution can be compared with densities measured for thick TiO$_2$ films (>1 μm), which became uniform with the thickness [48], presumably related to the reduction of the column growth competition and their extinction.

3.2. Second Series: Sputtering Pressure from 2.5×10^{-3} to 15×10^{-3} mbar

3.2.1. Morphology and Structure

The columnar microstructure and surface morphology of 400 nm thick W thin films were also influenced by the argon sputtering pressure, as shown in Figure 6.

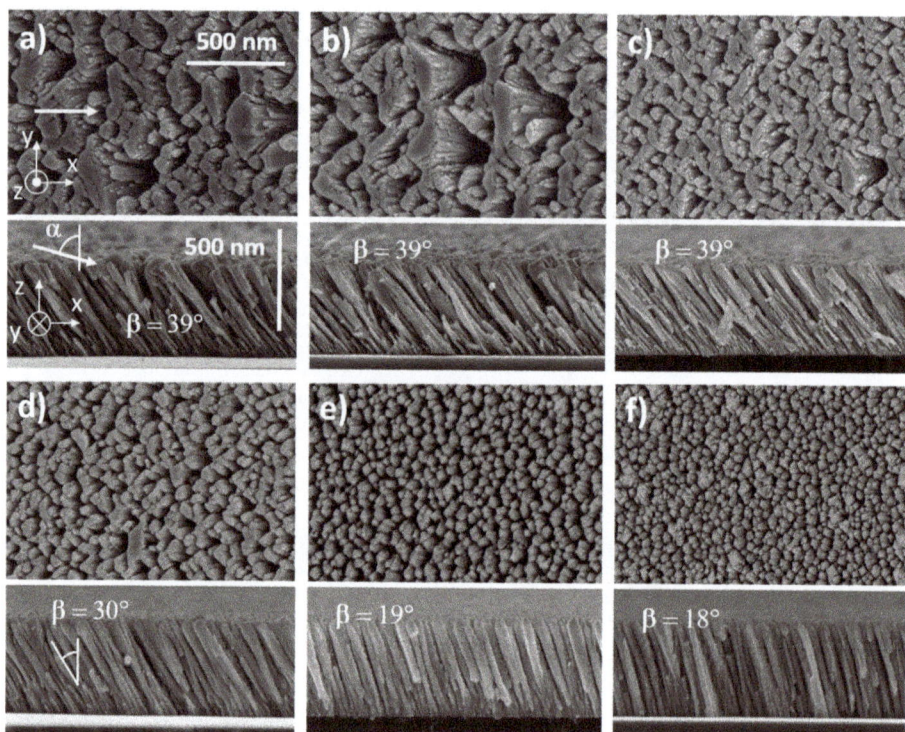

Figure 6. Top and cross-section observations by SEM of W films sputter-deposited with a substrate angle $\alpha = 80°$. The deposition time was set to get a constant film thickness of 400 nm. White arrows indicate the direction of incoming particle flux. Column angle β changed from $39° \pm 2°$ to $18° \pm 2°$ as the argon sputtering pressure increased from 2.5×10^{-3} to 15×10^{-3} mbar: (**a**) 2.5; (**b**) 3.5; (**c**) 4.0; (**d**) 6.0; I 10, and (**f**) 15×10^{-3} mbar.

For the lowest pressures (2.5×10^{-3} to 4.0×10^{-3} mbar in Figure 6a–6c, respectively), films exhibited a similar surface morphology, i.e., an elongated-shape of the columnar cross-section in the direction perpendicular to the particle flux (y-axis). Such anisotropic microstructure became even more marked as the argon sputtering pressure reduced with column widths reaching more than 500 nm for the largest spaces between columns of a few hundred nanometers in the x-direction. It is also interesting to note that this range of pressures gave rise to the highest column angle with β around 39° ± 2°, which was lower than the substrate angle $\alpha = 80°$, as expected in GLAD deposition. From the top-view images, the column apex in front of the particle flux exhibited a smoother and more abrupt edge than the opposite side (in the shadowing region), which showed a serrated edge and a fibrous feature. This difference of morphological microstructure between the opposite sides of the columns is related to the ballistic character of W sputtered atoms. For the lowest argon sputtering pressures, W atoms have a ballistic behavior. The compact part on the column side facing the flux is due to the energy transfer of W atoms. They impinged on the column apex and were abruptly stopped leading to a dense material. On the opposite side, located in the shadowing zone, a few parts of W atoms arrived with a grazing incidence. They hit the column apex close to the shadowing zone and induced atomic mobility processes in the direction of the particle flux [36]. Columns were then formed by a dense zone on one side, whereas a fibrous and porous one was produced on the opposite side.

Increasing the argon sputtering pressure up to 15×10^{-3} mbar (Figure 6f) led to a less anisotropic microstructure. Columns showed a more isotropic cross-section, and voids between columns reduced but were still relevant. In addition, the column angle decreased and reached $18° \pm 2°$ for this highest argon sputtering pressure. This change in microstructure has to be related to the thermalization of sputtered W atoms. Their energy decreased and the flux became scattered as the pressure rose. Taking into account our target-to-substrate distance (65 mm) and assuming that the calculated W atoms mean free path decreased from 7.2 cm down to 1.2 cm as the argon sputtering pressure changed from 2.5×10^{-3} to 15×10^{-3} mbar, respectively [49], direction and energy of W atoms impinging on the growing film were both modified. As determined by Westwood [50], the number of collisions required to thermalize a sputtered particle in an argon plasma is about 10. One can claim that for our range of pressures, W sputtered atoms changed from a ballistic to a thermalized characteristic mainly. From Barranco et al. [51], the thermalization degree Ξ and deposition angle α allow defining a microstructure phase map illustrating different kinds of microstructures in sputter-deposited thin films. A γ- to δ-type microstructural evolution is suggested for our W thin films as the argon sputtering pressure rose, i.e., some well-defined and isolated tilted columns at low thermalization degree (directional particle flux and low pressure), whereas a vertical and coalescent column-like structure with a high density of micro and mesopores occluded in the material for the highest one.

From XRD analyses (Figure 7), W films prepared at the lowest argon sputtering pressure (2.5×10^{-3} mbar) displayed better crystallinity.

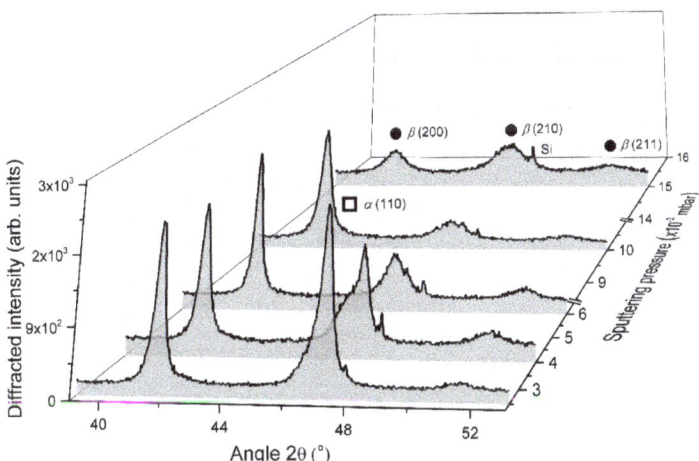

Figure 7. XRD patterns of 400 nm thick W thin films sputter-deposited on (100) Si substrate with an angle $\alpha = 80°$, and for an argon sputtering pressure changing from 2.5×10^{-3} to 15×10^{-3} mbar. Diffracted signals show the occurrence of α and β phases (π and λ, respectively).

An α and β phase mixture were clearly recorded with strong diffracted signals corresponding to α (110) and β (200) planes at $2\theta = 47.12°$ and $41.76°$, respectively. As previously reported in Section 3.1.1, it is interesting to note that the peak at $2\theta = 47.12°$ related to the α phase was not symmetric with a shoulder on the low-angle side, which is assigned to the diffraction by the β (210) planes. Increasing the argon sputtering pressure gave rise to lower diffracted signals for both phases. The shoulder-peak previously noticed became even more substantial, and for pressures higher than 6.0×10^{-3} mbar, only signals connected to the β phase were clearly recorded. Peaks became broad and weak for the highest argon sputtering pressure of 15×10^{-3} mbar. As the pressure increased from 2.5×10^{-3} to 15×10^{-3} mbar, the α phase vanished and crystal size of the β phase reduced (from Scherrer formula and assuming β (200) peak) from 28 to 10 nm, respectively. This decrease in crystallinity with the vanishing of the α phase vs. pressure well agrees with previous studies reporting W thin film growth by conventional sputtering [49,52]. A high argon sputtering pressure increases the probability of collision between sputtered atoms traveling toward the substrate and argon atoms and ions (thermalization prevails). The mean free path of W sputtered atoms (well below the target-to-substrate distance of 65 mm), as well as their energy, is then reduced. As a result, the W particle flux tends to be less directional and less energetic. As suggested by Vüllers and Spolenak [49], the reduced energy of incoming W atoms favors low mobility adsorption. Other phenomena, such as implantation and adsorption of high mobility atoms which prevail at low working pressure, become negligible. In addition, migration of W atoms to the α phase growth sites is hindered, and the reduction of W mobility by adsorbed argon atoms is favored [53,54]. Since the formation of the β phase is often accompanied by a porous architecture, some growing defects, and a secondary growth phenomenon [42,49,55], the occurrence of such crystallographic structure is more sensitive to environmental interactions such as residual oxygen, which prevents the long-range order and development of α phase.

3.2.2. Electrical Resistivity

Argon sputtering pressure did not only impact on the morphology and crystal structure of W GLAD thin films but it also influenced their electrical and anisotropic resistivity, as shown in Figure 8. A continuous and gradual increase in resistivity from $\rho_{300K} = 5.7 \times 10^{-6}$ to 5.6×10^{-5} Ω m was measured when the pressure changed from 2.5×10^{-3} to 15×10^{-3} mbar, respectively.

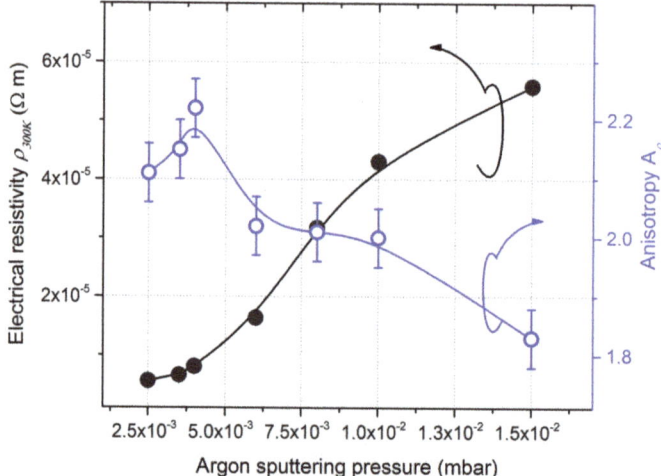

Figure 8. DC electrical resistivity and anisotropic resistivity measured at room temperature (300 K) vs. argon sputtering pressure of 400 nm thick W GLAD thin films prepared on a glass substrate with a deposition angle α = 80°.

Whatever the argon sputtering pressure, the films' resistivity was again two to three orders of magnitude higher than the W bulk value ($\rho_{300K} = 5.4 \times 10^{-8}$ Ω m [38]). This difference is commonly attributed to the grain boundary scattering of free electrons. XRD analyses showed a polycrystalline structure and the best crystallinity for films prepared with the lowest argon sputtering pressure of 2.5×10^{-3} mbar (Figure 7). For such a pressure, the coexistence of α and β phases was recorded with a crystal size of 22 and 23 nm, respectively (calculated from Scherrer formula and after deconvolution of diffracted signal related to β (210) and α (110) peaks). These values were higher than the electron mean free path in W, which is about 16 nm [56]. It means that the resistivity of GLAD W films produced at low argon sputtering pressure cannot be completely assigned to the electron scattering at the grain boundaries. The high voided microstructure (usually obtained in GLAD films) and large spaces between the tilted columns clearly observed from SEM images (Figure 6a) rather contributed to the high resistivity.

A substantial increase in resistivity was measured for an argon sputtering pressure higher than 6.0×10^{-3} mbar. This pressure range has to be correlated with the gradual evolution of the films' microstructure (Figure 6), vanishing of the α phase and smooth decrease in the crystal size of the β phase down to a few nanometers for the highest pressure (Figure 7). Despite the shorter spaces between tilted columns as the pressure rose, voids still remained and contributed to the electron scattering. In addition, XRD patterns showed a broadening of diffracted signals related to the β phase, which means a crystal size lower than the electron mean free path. As a result, the film's resistivity was dominated by characteristics of the remaining β phase, the latter being more resistive than the α phase.

A significant difference in electrical resistivity was measured following x- and y-directions (determined from the Bierwagen method [23]). This difference between ρ_x and ρ_y is once again related to the microstructure of W thin films. It was clearly demonstrated that GLAD deposition of columnar films with incident angles higher than 60° (critical-like angle) produced significant structural and uniaxial anisotropy in the substrate plane induced by the shadowing effect [57]. Thus, an anisotropic distribution of the grain boundary potential barrier heights was formed in the direction of the incoming atoms (x-axis), especially at low argon sputtering pressure due to the high directional behavior of the particle flux. W nuclei grew with a connection to each other by chains perpendicular to the direction of the shadowing effect (y-axis). The columnar growth exhibited an elliptical-shape cross-section, which became less marked as the pressure rises. Therefore, this difference of electrical resistivity in the orthogonal axes was less and less noticeable from 6.0×10^{-3} mbar, which is the pressure range corresponding to the formation of more isotropic cross-section and narrow columns. This change in microstructure is also related to the thermalization of sputtered W atoms (their energy decreases and the flux becomes less directional). Anisotropy was above 2.1 for the lowest argon sputtering pressures and progressively reduced to about 1.8 at 15×10^{-3} mbar. This A_ρ value was still high despite the isotropic columnar microstructure viewed from SEM pictures (Figure 6f). This means that electron scattering remained preferential in the direction to the particle flux (x-axis) and, thus a structural anisotropy remained as we can see in Figure 6e,f, where the density of columns was high, and they tended to keep bundled following the x-direction.

3.2.3. Elastic Wave Propagation

The elastic properties of W films also depend on the argon sputtering pressure. Figure 9 illustrates the evolution of the group velocities and anisotropy of the pseudo-Rayleigh wave as the pressure changed from 2.5×10^{-3} to 15×10^{-3} mbar.

For the lowest pressures (2.5×10^{-3} to 4.0×10^{-3} mbar), the sputtered vapor was strongly directional, which favors shadowing conditions. The lateral growth allowed neighboring columns to touch and chain together, whereas shadowing prevented columns merging along the x-direction. This produced a preferential bundling of the columnar microstructure in the y-direction [13]. Hence, the velocity was significantly lower following the x-direction, which led to an important elastic anisotropic coefficient (A_v = 1.8 to 2). Both velocities in the x- and y-directions increased with argon sputtering pressure. This evolution was correlated with the microstructural changes in the films and the reduction of porosity (Table 2).

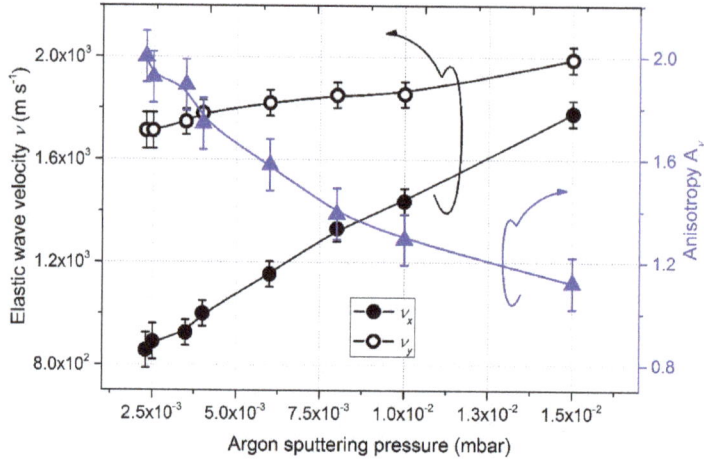

Figure 9. Group velocities of pseudo-Rayleigh waves along x and y-axes and related anisotropic coefficient as a function of the argon sputtering pressure.

Table 2. Calculated porosity as a function of the argon sputtering pressure.

Pressure P_{Ar} ($\times 10^{-3}$ mbar)	2.5	3.5	4.0	6.0	8.0	10.0	15.0
Global porosity π ($\pm 10\%$ of the bulk)	68	65	62	59	55	53	46

Increasing the argon sputtering pressure up to 15×10^{-3} mbar (Figure 6f) gave rise to a less anisotropic microstructure. Columns showed a more isotropic cross-section and voids between them became smaller. As previously mentioned, this change in microstructure is related to the thermalization of sputtered W atoms. A further increase in pressure induced less inclined columns as the lack of vapor directionality effectively reproduced an isotropic deposition geometry. Nonetheless, a previous study [24] showed that the column tilt angle β has hardly any influence on velocities, and anisotropic velocity is mainly connected to the holes/columns geometry, especially their form factor. Due to the rather circular shape of the column cross-sections, x and y-velocities were similar with $v_x = 1800$ m s^{-1} and $v_y = 2000$ m s^{-1}, which means an anisotropy A_v around 1.2. These results are consistent with an earlier publication where no anisotropy on the pseudo-Rayleigh waves propagating in gold films deposited by GLAD are reported (SEM observations similarly showed nearly circular column cross-sections) [46].

As reported in Table 2, the global porosity rapidly decreased as the argon sputtering pressure increased from $\pi = 68 \pm 10\%$ for 2.5×10^{-3} mbar to $46 \pm 10\%$ of the bulk for 15×10^{-3} mbar. This result comes from the decrease in the inter-columnar space, leading to the densification of the film. Indeed, the high voided microstructure and large spaces between the tilted columns (obtained for the lowest pressures) clearly viewed by SEM analyses (Figure 6a–c) contributed to the global porosity. Although an increase in argon sputtering pressure usually leads to much more porous thin films prepared by conventional sputtering, this reverse trend has previously been reported in Liang et al. investigations for Mg films deposited by the GLAD method [58]. The authors also reached the same conclusion, i.e., a decrease in the film porosity due to shorter intercolumnar spaces obtained at high pressure.

4. Conclusions

Tungsten thin films were prepared by DC magnetron sputtering by the GLAD technique. A constant deposition angle α = 80° was used for all depositions. For a first series of films, a constant argon sputtering pressure of 4.0×10^{-3} mbar was used, whereas the film's thickness was progressively changed from 50 to 1000 nm. An anisotropic microstructure was developed as the film's thickness increased. The column angle was kept constant around 39° whatever the film's thickness and columns became more and more asymmetric with elongated columnar cross-sections perpendicular to the incoming particle flux. This anisotropic microstructure was correlated with the ballistic character of sputtered particles (high directional flux), and the shadowing effect was effective from the early growing stage. DC electrical resistivity was progressively reduced as a function of the film's thickness due to an increase in the grain size and the occurrence of the α-W phase. Resistivity, as well as surface elastic wave velocity, were reduced as the film's thickness increases due to a more voided architecture (an increase in the micro- and nanoporosity). They also both exhibited an enhanced anisotropy up to 450 nm thick, which was correlated with a thickness range corresponding to a better-defined anisotropic microstructure.

For the second series, deposition time was adjusted to get a constant film thickness of 400 nm, and the argon sputtering pressure was systematically varied from 2.5×10^{-3} to 15×10^{-3} mbar. The elongated shape of the columnar cross-section produced with the lowest pressures in the direction perpendicular to the particle flux progressively transformed to less tilted and narrow columns with a more isotropic cross-section for the highest pressures. This morphological evolution was assigned to the predominance of thermalized tungsten atoms, more dispersive and less energetic tungsten atoms impinging on the growing films. Increasing the argon sputtering pressure, the α phase completely vanished, and films became poorly crystallized with the growth of a nano-crystallized β phase. This decrease in crystallinity was associated with an enhancement of the electrical resistivity, whereas surface elastic wave velocity was gradually increased due to the development of less spaced columns (decrease in porosity). The columnar microstructure became more homogeneous at high pressure and induced a more isotropic behavior of electronic conduction and elastic wave propagation in columnar tungsten thin films.

Author Contributions: All authors have read and agreed to the published version of the manuscript. Data curation, A.C. and R.E.B.; Formal analysis, V.P.; Funding acquisition, P.V. and N.M.; Investigation, A.M. and S.E. All authors have read and agreed to the published version of the manuscript.

Funding: This work was supported by the EIPHI Graduate School (contract ANR-17-EURE-0002), the Region of Bourgogne Franche-Comté and the French RENATECH network. This research was also sponsored by the French National Agency for Research (ANR) funds through the PHEMTO project.

Acknowledgments: J.M. Cote is acknowledging for technical assistance.

Conflicts of Interest: The authors declare no conflict of interest. The funders had no role in the design of the study; in the collection, analyses, or interpretation of data; in the writing of the manuscript, or in the decision to publish the results.

References

1. Baranovski, S. *Charge Transport in Disordered Solids with Applications in Electronics*; John Wiley & Sons Ltd.: Hoboken, NJ, USA, 2006.
2. Wu, T.T.; Huang, Z.G.; Lin, S. Surface and bulk acoustic waves in two-dimensional phononic crystal consisting of materials with general anisotropy. *Phys. Rev. B* **2004**, *69*, 094301. [CrossRef]
3. Lundstrom, M. *Fundamentals of Carrier Transport*; Cambridge University Press: Cambridge, UK, 2000.
4. Guo, L.; Ren, Y.; Kong, L.Y.; Chim, W.K.; Chiam, S.Y. Ordered fragmentation of oxide thin films at submicron scale. *Nat. Commun.* **2016**, *7*, 13148. [CrossRef]
5. Luo, Z.; Maassen, J.; Deng, X.; Du, Y.; Garrelts, R.P.; Lundstrom, M.S.; Ye, P.D.; Xu, X. Anisotropic in-plane thermal conductivity observed in few-layer black phosphorus. *Nat. Commun.* **2015**, *6*, 8572–8578. [CrossRef] [PubMed]
6. Belardini, A.; Centini, M.; Leahu, G.; Hooper, D.C.; Voti, R.L.; Fazio, E.; Haus, J.W.; Sarangan, A.; Valev, V.K.; Sibilia, C. Chiral light intrinsically couples to extrinsic/pseudo-chiral metasurfaces made of tilted gold nanowires. *Sci. Rep.* **2016**, *6*, 31796–31799. [CrossRef] [PubMed]

7. Zhu, G.; Liu, J.; Zheng, Q.; Zhang, R.; Li, D.; Banerjee, D.; Cahill, D.G. Tuning thermal conductivity in molybdenum disulfide by electrochemical intercalation. *Nat. Commun.* **2016**, *7*, 13211–13219. [CrossRef] [PubMed]
8. Probst, P.T.; Sekar, S.; König, T.A.F.; Formanek, P.; Decher, G.; Fery, A.; Pauly, M. Highly oriented nanowire thin films with anisotropic optical properties driven by the simultaneous influence of surface templating and shear forces. *ACS Appl. Mater. Interfaces* **2018**, *10*, 3046–3057. [CrossRef] [PubMed]
9. Vergara, J.; Favieres, C.; Magen, C.; Teresa, J.M.; Ibarra, M.R.; Madurga, V. Structurally oriented nano-sheets in Co thin films: Changing their anisotropic physical properties by thermally induced relaxation. *Materials* **2017**, *10*, 1390. [CrossRef]
10. Mohanty, B.; Ivanoff, T.A.; Alagoa, A.S.; Karabacak, T.; Zou, M. Study of the anisotropic frictional and deformation behavior of surfaces textured with silver nanorods. *Tribol. Int.* **2015**, *92*, 439–445. [CrossRef]
11. Sun, Y.; Xiao, X.; Xu, G.; Dong, G.; Chai, G.; Zhang, H.; Liu, P.; Zhu, H.; Zhan, Y. Anisotropic vanadium dioxide sculptured thin films with superior thermochromic properties. *Sci. Rep.* **2013**, *3*, 2756. [CrossRef]
12. Krausch, G.; Magerle, R. Nanostructured thin films via self-assembly of block copolymers. *Adv. Mater.* **2002**, *14*, 1579–1583. [CrossRef]
13. Hawkeye, M.; Taschuk, M.T.; Brett, M.J. *Glancing Angle Deposition of Thin Films—Engineering the Nanoscale*; Wiley: Hoboken, NJ, USA, 2014.
14. Plawsky, J.L.; Kim, J.K.; Schubert, E.F. Engineered nanoporous and nanostructured films. *Mater. Today* **2009**, *12*, 36–45. [CrossRef]
15. Grüner, C.; Liedtke, S.; Bauer, J.; Mayr, S.G.; Rauschenbach, N. Morphology of thin films formed by oblique physical vapor deposition. *ACS Nano Mater.* **2018**, *1*, 1370–1376. [CrossRef]
16. Robbie, K.; Brett, M.J.; Lakhtakia, A. Chiral sculptured thin films. *Nature* **1996**, *384*, 616–617. [CrossRef]
17. Song, C.; Larsen, G.K.; Zhao, Y. Anisotropic resistivity of tilted silver nanorod arrays: Experiments and modeling. *Appl. Phys. Lett.* **2013**, *102*, 233101–233104. [CrossRef]
18. Charles, C.; Martin, N.; Devel, M. Optical properties of nanostructured WO_3 thin films by glancing angle deposition: Comparison between experiment and simulation. *Surf. Coat. Technol.* **2015**, *276*, 136–140. [CrossRef]
19. Martin, N.; Sauget, J.; Nyberg, T. Anisotropic electrical resistivity during annealing of oriented columnar titanium films. *Mater. Lett.* **2013**, *105*, 20–23. [CrossRef]
20. Garcia-Martin, J.M.; Alvarez, R.; Romero-Gomez, P.; Cebollada, E.; Palmero, A. Tilt angle control of nanocolumns grown by glancing angle sputtering of variable argon pressures. *Appl. Phys. Lett.* **2010**, *97*, 173103–173113. [CrossRef]
21. Mukherjee, S.; Gall, D. Structure zone model for extreme shadowing conditions. *Thin Solid Films* **2013**, *527*, 158–163. [CrossRef]
22. Lintymer, J.; Gavoille, J.; Martin, N.; Takadoum, J. Glancing angle deposition to modify microstructure and properties of chromium thin films sputter deposited. *Surf. Coat. Technol.* **2003**, *174*, 316–323. [CrossRef]
23. Bierwagen, O.; Pomraenke, R.; Eilers, S.; Masselink, W.T. Mobility and carrier density in materials with anisotropic conductivity revealed by van der Pauw measurements. *Phys. Rev. B* **2004**, *70*, 165307. [CrossRef]
24. Coffy, E.; Dodane, G.; Euphrasie, S.; Mosset, A.; Vairac, P.; Martin, N.; Baida, H.; Rampnoux, J.M.; Dilhaire, S. Anisotropic propagation imaging of elastic waves in oriented columnar thin films. *J. Phys. D: Appl. Phys.* **2017**, *50*, 484005–484008. [CrossRef]
25. Royer, D.; Dieulesaint, D. *Ondes élastiques dans les Solides II: Génération, Interaction Acoustico-Optique, Applications*; Elsevier Masson: Paris, France, 1996.
26. Biot, M.A. Theory of propagation of elastic waves in a fluid-saturated porous solid. II. Higher frequency range. *J. Acoust. Soc. Am.* **1956**, *28*, 179–191. [CrossRef]
27. Da Fonseca, R.J.M.; Saurel, J.M.; Foucaran, A.; Camassel, J.; Massone, E.; Taliercio, T.; Boumaiza, Y. Acoustic investigation of porous silicon layers. *J. Mater. Sci.* **1995**, *30*, 35–39. [CrossRef]
28. Aliev, G.N.; Goller, B.; Snow, P.A. Elastic properties of porous silicon studied by acoustic transmission spectroscopy. *J. Appl. Phys.* **2011**, *110*, 43534–43541. [CrossRef]
29. Reinhardt, A.; Snow, P.A. Theoretical study of acoustic band-gap structures made of porous silicon. *Phys. Status Solidi* **2007**, *204*, 1528–1535. [CrossRef]

30. El Beainou, R.; Martin, N.; Potin, V.; Pedrosa, P.; Arab Pour Yazdi, M.; Billard, A. Correlation between structure and electrical resistivity of W-Cu thin films prepared by GLAD co-sputtering. *Surf. Coat. Technol.* **2017**, *313*, 1–7. [CrossRef]
31. Karabacak, T.; Singh, J.P.; Zhao, Y.P.; Wang, G.C.; Lu, T.M. Scaling during shadowing growth of isolated nanocolumns. *Phys. Rev. B* **2003**, *68*, 125408. [CrossRef]
32. Alvarez, R.; Garcia-Martin, J.M.; Garcia-Valenzuela, A.; Macias-Montero, M.; Ferrer, F.J.; Santiso, J.; Rico, V.; Cotrino, J.; Gonzalez-Elipe, A.R.; Palmero, A. Nanostructured Ti thin films by magnetron sputtering at oblique angles. *J. Phys. D: Appl. Phys.* **2016**, *49*, 045303. [CrossRef]
33. Maillé, L.; Sant, C.; Le Paven-Thivet, C.; Legrand-Buscema, C.; Garnier, P. Structure and morphological study of nanometer W and W_3O thin films. *Thin Solid Films* **2003**, *428*, 237–241. [CrossRef]
34. Choi, D. Phase transformation in thin tungsten films during sputter deposition. *Microelectron. Eng.* **2017**, *183*, 19–22. [CrossRef]
35. Petroff, P.; Sheng, T.T.; Sinha, A.K.; Rozgonyi, G.A.; Alexander, F.B. Microstructure, growth, resistivity, and stresses in thin tungsten films deposited by rf sputtering. *J. Appl. Phys.* **1973**, *44*, 2545–2554. [CrossRef]
36. El Beainou, R.; Salut, R.; Robert, L.; Cote, J.M.; Potin, V.; Martin, N. Anisotropic conductivity enhancement in inclined W-Cu columnar films. *Mater. Lett.* **2018**, *232*, 126–129. [CrossRef]
37. Kaidatzis, A.; Psycharis, V.; Mergia, K.; Niarchos, D. Annealing effects on the structural and electrical properties of sputtered tungsten thin films. *Thin Solid Films* **2016**, *619*, 61–67. [CrossRef]
38. Lide, D.R. *CRC Handbook of Chemistry and Physics*; CRC Press: Boca Raton FL, USA, 2005.
39. Besnard, A.; Martin, N.; Carpentier, L.; Gallas, B. A theoretical model for the electrical properties of chromium thin films sputter deposited at oblique incidence. *J. Phys. D: Appl. Phys.* **2011**, *44*, 215301–215308. [CrossRef]
40. Kasap, S.; Capper, P. *Springer Handbook of Electronic and Photonic Materials*; Springer: New York, NY, USA, 2007.
41. Petroff, P.M.; Reed, W.A. Resistivity behavior and phase transformations in β-W thin films. *Thin Solid Films* **1974**, *21*, 73–81. [CrossRef]
42. Choi, D.; Wang, B.; Chung, S.; Liu, X.; Darbal, A.; Wise, A.; Nuhfer, N.T.; Barmak, K. Phase, grain structure, stress, and resistivity of sputter-deposited tungsten films. *J. Vac. Sci. Technol. A* **2011**, *29*, 051512. [CrossRef]
43. Amassian, A.; Kaminska, K.; Suzuki, M.; Martinu, L.; Robbie, K. Onset of shadowing-dominated growth in glancing angle deposition. *Appl. Phys. Lett.* **2006**, *91*, 173114. [CrossRef]
44. Drechsler, M.; Blackford, B.; Putnam, A.; Jericho, M. A measurement of a surface self-diffusion coefficient by scanning tunneling microscopy. *J. Phys. Colloq.* **1989**, *50*, 223–228. [CrossRef]
45. Vick, D.; Brett, M.J. Conduction anisotropy in porous thin films with chevron microstructures. *J. Vac. Sci. Technol. A* **2006**, *24*, 156–164. [CrossRef]
46. El Beainou, R.; Chargui, A.; Pedrosa, P.; Mosset, A.; Euphrasie, S.; Vairac, P.; Martin, N. Electrical conductivity and elastic wave propagation anisotropy in glancing angle deposited tungsten and gold films. *Appl. Surf. Sci.* **2019**, *475*, 606–614. [CrossRef]
47. Mason, W.P.; Thurston, R.N. *Physical Acoustics: Principles and Methods*; Academic Press: New York, NY, USA, 1970.
48. Krause, K.M.; Taschuk, M.T.; Harris, K.D.; Rider, D.A.; Wakefield, N.G.; Sit, J.C.; Buriak, J.M.; Thommes, M.; Brett, M.J. Surface area characterization of obliquely deposited metal oxide nanostructured thin films. *Langmuir* **2010**, *26*, 4368–4376. [CrossRef] [PubMed]
49. Vüllers, F.T.N.; Spolenak, R. Alpha- vs. beta-W nanocrystalline tungsten thin films: A comprehensive study of sputter parameters and resulting materials' properties. *Thin Solid Films* **2015**, *577*, 26–34. [CrossRef]
50. Weswtood, W.D. Calculation of deposition rates in diode sputtering systems. *J. Vac. Sci. Technol.* **1978**, *15*, 1–9. [CrossRef]
51. Barranco, A.; Borras, A.; Gonzalez-Elipe, A.R.; Palmero, A. Perspectives on oblique angle deposition of thin films: From fundamentals to devices. *Prog. Mater. Sci.* **2016**, *76*, 59–153. [CrossRef]
52. Djerdi, I.; Tonejc, A.M.; Tonejc, A.; Radic, N. XRD lines profile analysis of tungsten thin films. *Vacuum* **2005**, *80*, 151–158. [CrossRef]
53. Shen, Y.G.; Mai, Y.W.; Zhang, Q.C.; McKenzie, D.R.; McFall, W.D.; McBride, W.E. Residual stress, microstructure, and structure of tungsten thin films deposited by magnetron sputtering. *J. Appl. Phys.* **2000**, *87*, 177–187. [CrossRef]

54. Thornton, J.A. Influence of apparatus geometry and deposition conditions on the structure and topography of thick sputtered coatings. *J. Vac. Sci. Technol.* **1974**, *11*, 666–670. [CrossRef]
55. O'Keefe, M.J.; Grant, J.T. Phase transformation of sputter deposited tungsten thin films with A-15 structure. *J. Appl. Phys.* **1996**, *79*, 9134–9141. [CrossRef]
56. Gall, D. Electron mean free path in elemental metals. *J. Appl. Phys.* **2016**, *119*, 085101. [CrossRef]
57. Siad, A.; Besnard, A.; Nouveau, C.; Jacquet, P. Critical angles in DC magnetron glad thin films. *Vacuum* **2016**, *131*, 305–311. [CrossRef]
58. Liang, H.; Geng, X.; Li, W.; Panepinto, A.; Thiry, D.; Chen, M.; Snyders, R. Experimental and modeling study of the fabrication of Mg nano-sculpted films by magnetron sputtering combined with glancing angle deposition. *Coatings* **2019**, *9*, 361. [CrossRef]

© 2020 by the authors. Licensee MDPI, Basel, Switzerland. This article is an open access article distributed under the terms and conditions of the Creative Commons Attribution (CC BY) license (http://creativecommons.org/licenses/by/4.0/).

Article

Spatially Resolved Optoelectronic Properties of Al-Doped Zinc Oxide Thin Films Deposited by Radio-Frequency Magnetron Plasma Sputtering Without Substrate Heating

Eugen Stamate

National Centre for Nano Fabrication and Characterization, Technical University of Denmark, Ørsteds Plads, 2800 Kongens Lyngby, Denmark; eust@dtu.dk

Received: 29 November 2019; Accepted: 17 December 2019; Published: 19 December 2019

Abstract: Transparent and conducting thin films were deposited on soda lime glass by RF magnetron sputtering without intentional substrate heating using an aluminum doped zinc oxide target of 2 inch in diameter. The sheet resistance, film thickness, resistivity, averaged transmittance and energy band gaps were measured with 2 mm spatial resolution for different target-to-substrate distances, discharge pressures and powers. Hall mobility, carrier concentration, SEM and XRD were performed with a 3 mm spatial resolution. The results reveal a very narrow range of parameters that can lead to reasonable resistivity values while the transmittance is much less sensitive and less correlated with the already well-documented negative effects caused by a higher concentration of oxygen negative ions and atomic oxygen at the erosion tracks. A possible route to improve the thin film properties requires the need to reduce the oxygen negative ion energy and investigate the growth mechanism in correlation with spatial distribution of thin film properties and plasma parameters.

Keywords: transparent conducting oxides; aluminum doped zinc oxide; magnetron plasma sputtering

1. Introduction

Transparent and conductive thin films are important for a large number of applications, including but not limited to: touch screens, solar cells, smart windows (low-e, chromogenic devices) and light emitting diodes [1–5]. Oxides doped with metals, generically known as transparent conductive oxides (TCO) are successfully used nowadays, with indium tin oxide (ITO) being the best material, with a resistivity of around 10^{-4} Ωcm and transmittance above 88% [1]. However, the high demand for large area applications, coupled with the reduced abundance of indium, limits market penetration for very large area applications, such as low-e windows and solar cells [3,4]. This motivation sustains intensive research on alternative materials, with aluminum-doped zinc oxide (AZO) being one of the most promising choice due to the high abundance of Zn and Al [2]. For example, cost effective solar cells based on Cu (In,Ga)Se$_2$ (CIGS) and Cu$_2$ZnSnS$_4$ (CZTS) absorbers have been fabricated with a TCO based on AZO [6,7]. There are several methods used to deposit AZO, including physical vapor deposition (under various operation conditions for magnetron sputtering, such as radio-frequency [8–20], medium-frequency [8,21–26], DC [8,16,22,26–37], pulsed DC [38], high power impulse [39,40], ion beam assisted [41], chemical vapor deposition [1,5] and other chemical methods such as spin coating and sol gel [2,4]. Among them, magnetron plasma sputtering has been successfully used to deposit ITO on large area substrates (up to 15 m^2) and is also regarded as a viable and cost effective solution for AZO [1–5]. However, the resistivity of AZO thin films is about 5 to 10 times higher than that of ITO, with better values only for limited locations on the substrate [2,5]. The main reason for this is the electronegativity of oxygen that easily forms negative ions by attaching low-energy electrons emitted from the target by secondary emission or generated by plasma [2,5]. Since the sputtering target builds up a negative bias (positive ions produce the sputtering after being accelerated in a thin space charge layer named the plasma sheath), the negative ions [42] are accelerated over the sheath towards the substrate, and assist the film growth with energies distributed from 0 to 500 eV for operation in DC and 0–300 eV for operation in radio-frequency discharge, as recently reported by Ellmer et al. [43,44]. The presence of permanent magnets behind the target, so as to produce a high-density plasma close to the surface, results in a non-uniform erosion. This non-uniformity is correlated with the radial distribution of the negative ions that is eventually mirrored on the substrate [45]. At the same time, one expects a non-uniform distribution of the oxygen released from the target (mainly for short target-to-substrate distances) which can also influence the thin film growth [5]. Up to date, both energetic negative ions and oxygen distribution are considered as the main reasons responsible for the poor optoelectronic properties of AZO over the substrate surface [2,5,45]. AZO deposited by magnetron plasma sputtering was, and yet is, intensively studied, with a large number of works reporting resistivity values in the range of 10^{-4} Ωcm only for small substrate areas or for films with transmittance below 80% [1–5]. The lowest resistivity for AZO, of 8.54×10^{-5} Ωcm, was reported by pulsed laser deposition over a non-specified area [46]. However, a critical investigation reveals that further improvement of AZO properties could only be possible by simultaneously achieving a resistivity below 3×10^{-4} Ωcm and a transmittance above 88%, for a substrate area comparable with the target area. If these optoelectronic properties are achieved without intentional substrate heating, then such a process can be used to coat heat sensitive substrates.

The aim of this work is to identify the best range of deposition parameters (pressure, power and target-to-substrate distance) and to provide a set of carefully measured spatial distribution profiles for optoelectronic properties (sheet resistance, resistivity, mobility, carrier concentration and transmittance) of AZO thin films that can be used as a reference for further studies so as to eventually understand the thin film growth mechanism.

2. Materials and Methods

Clean soda lime glass samples of 10×50 mm and 0.75 mm in thickness were used as substrates to deposit AZO thin films by a two-inch in diameter Zinc/Alumina (ZnO/Al$_2$O$_3$, 97/2 wt%) target (Kurt Lesker, Jefferson Hills, PA, USA) mounted on a TORUS® cathode powered in radio-frequency at 13.56 MHz (see Figure 1a).

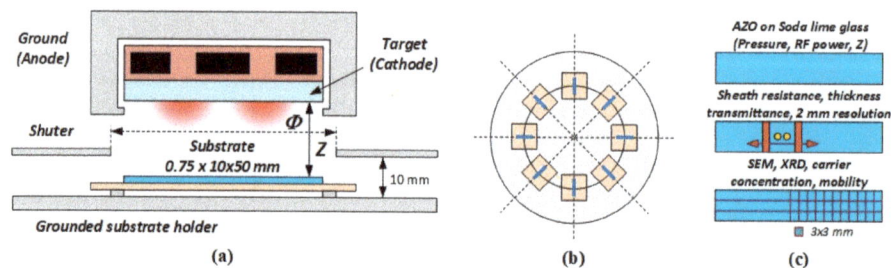

Figure 1. (a) Schematic of the experimental setup, (b) samples arrangement on the substrate holder, (c) sample characterization.

The vacuum chamber was large enough (50 cm in diameter) to accommodate 8 samples at the same time, placed on a large holder that could rotate so as to expose the samples one by one, to different discharge conditions (power, P_{RF}, pressure, p, and target-to-substrate distance, Z), without turning off the discharge (see Figure 1b). A large disk shutter was used to prevent deposition on samples during the time needed to adjust the discharge parameters. The deposition was done by aligning a Φ = 60 mm in diameter opening in the shutter disk between the target and the substrate at a constant distance of

10 mm between the opening and the substrate holder. The target-to- substrate distance was adjusted by translating the cathode upwards. The discharge pressure was varied in the range of 1.4 to 50 mTorr, the RF power from 10 up to 100 W and target-to-substrate distance from 25 up to 100 mm. There was no intentional heating of the substrate except for a temperature rise due to plasma exposure, which was less than 70 degrees (measured with a no-contact FTX-100-LUX + OSENSA Innovations, Burnaby, BC, Canada, temperature transmitter) after 1 h deposition time for all process parameters presented in this work. The sheath resistance was measured with a resolution of 2 mm using a four-probe system, configured to accommodate the substrate dimensions (see Figure 1c). The thin film thickness was measured with a thin film analyzer Filmetrics F20 (San Francisco, CA, USA) and confirmed by SEM with the same spatial resolution as the sheet resistance so as to enable the calculation of the AZO thin film resistivity profile. The transmittance spectra was measured using an Agilent Cary 100 UV-Vis photo-spectrometer (Santa Clara, Ca, USA), in steps of 2 mm, to provide the averaged transmittance in the range of 400 to 700 nm and the band gap energy from Tauc's plot. In the last steps of characterization, the samples were cut into 3 × 3 mm pieces as presented in Figure 1c and separate rows were used for SEM (Zeiss Merlin, Oberkochen, Germany), XRD and Hall effect measurement (ezHEMS from Nanomagnetics Instruments, Oxford, UK) of carrier concentration and mobility.

3. Results

The sheet resistance as a function of radial position ($r = 0$ at sample center) for Z = 45, and 65 mm, P_{RF} = 20 W and 60 min deposition time is presented in Figure 2a for p = 1.4 mTorr and (b) p = 3 mTorr, respectively, where 1.4 mTorr was the lowest pressure to sustain the plasma.

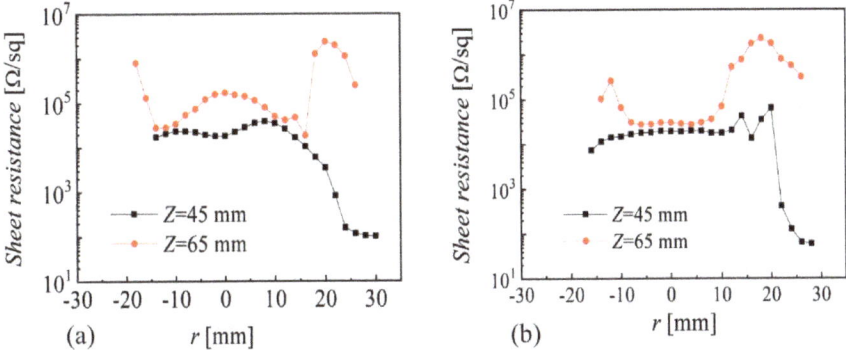

Figure 2. Sheet resistance as a function of radial position for (a) p = 1.4 mTorr and (b) p = 3 mTorr where P_{RF} = 20 W and deposition time 60 min.

Despite the small change in pressure, one can see a flattening of the sheet resistance for $-10 < r < 10$ mm when increasing the p from 1.4 to 3 mTorr. However, the most remarkable thing is the difference, of almost two orders of magnitude, for $20 < r < 30$ mm at both pressures, as well as the variation of the sheet resistance with more than one order of magnitude for Z = 45 mm at the locations corresponding to r = 10 mm and r = 25 mm. It is important to note that the sheet resistance profiles are symmetric with respect to r = 0 mm and on purpose we used a translation of 4 mm so as to be able to capture thin film properties for $25 < r < 30$ mm (close to the edge of the shutter). Figure 2a exhibits a good correlation with the erosion tracks for Z = 45 mm (two humps structure, for $r \sim -10$ mm and 10 mm respectively) while increasing Z to 65 mm gives a convolution, with a single hump for $-10 < r < 10$ mm. Such behavior was reported more than 20 years ago and was associated with the possible influence of negative ions or oxygen distribution [45]. Application wise, sheet resistance values above 100 Ω/sq were too high, and increasing Z led to even higher values, so that the target-to-substrate distance was further decreased. Figure 3a presents the sheath resistance and (b) the film thickness for different pressures and $Z = 35$ mm. In this case, the correlation with the erosion tracks is evident for $p \geq 3$ mTorr, with two humps that are getting closer by increasing p. The spatial distribution at $p = 1.4$ mTorr shows a strong central peak, revealing that the plasma discharge exhibits a torch-like profile, resulting from an inefficient plasma production by the magnetic field at low pressures. However, it is remarkable to see that a very small change in pressure (from 1.4 to 3 mTorr) has a significant effect on the plasma (revealed by the film thickness) and sheet resistance profile. Once again, one can notice very large variations in the sheet resistance, as the decrease from 10^4 Ω/sq to 350 Ω/sq, only by moving from $r = 5$ mm to $r = 17$ mm at $p = 9$ mTorr. The film thickness presented in Figure 3b helps one to understand that the torch-like discharge mode causes intensive re-sputtering on the sample surface for $-10 < r < 10$ mm, while higher pressures reveal almost parabolic profiles with some small shoulders correlated with the erosion track for 6 and 9 mTorr. A higher film thickness at 3 mTorr with respect to 1.4 mTorr suggests a significant change in plasma density within this narrow pressure change.

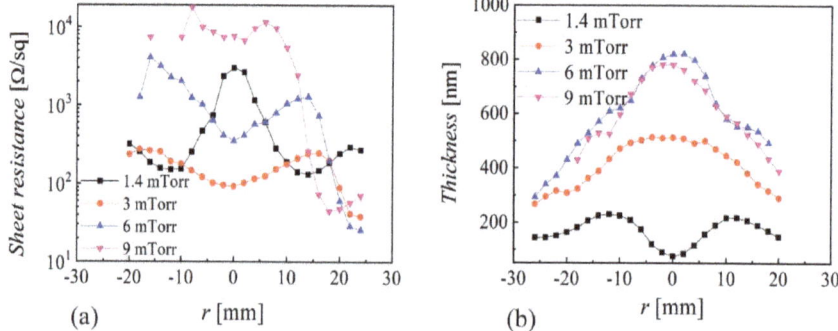

Figure 3. Spatial distribution of (**a**) sheet resistance and (**b**) film thickness for different pressures and $Z = 35$ mm and $P_{RF} = 20$ W.

The resistivity, Hall mobility and carrier concentration measured on 3×3 mm^2 samples (see Figure 1c) are presented in Figure 4 for $-5 < r < 25$ mm and $p = 1.4$ and 3 mTorr, with lowest resistivity and highest mobility and carrier concentration at the edge ($r = 23$ mm) of the sample deposited at 3 mTorr. The optical performance is characterized by the transmittance spectra in the visible range that have been measured between 250 and 900 nm. As example, the radial distribution of transmittance spectra from $r = -24$ mm to the sample center are presented in Figure 5 for $p = 3$ mTorr, $Z = 35$ mm and $P_{RF} = 20$ W, from which one can see the interference oscillations correlated with the film thickness. The averaged transmittance (400 to 700 nm) for 1.4 mTorr and 3 mTorr is presented in Figure 6a for $Z = 35$ mm and (b) for $Z = 45$ mm, where an obvious correlation with the erosion track can be seen only for the sample at 3 mTorr and $Z = 45$ mm. All of the values are above 87%, even reaching above 93% over the whole sample deposited at 1 mTorr and $Z = 45$ mm, thus revealing that the main challenge is to reduce the resistivity.

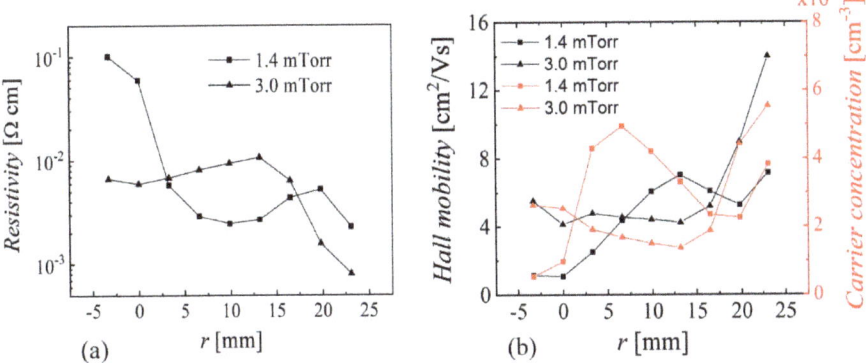

Figure 4. Spatial distribution of (**a**) resistivity and (**b**) mobility and carrier concentration for samples deposited at 1.4 and 3 mTorr, $Z = 35$ mm and $P_{RF} = 20$ W.

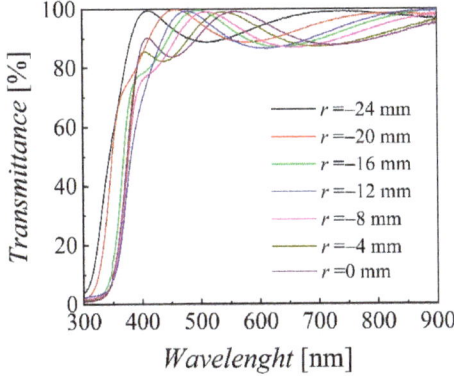

Figure 5. Transmittance spectra at different radial locations for $p = 3$ mTorr, $Z = 35$ mm and $P_{RF} = 20$ W.

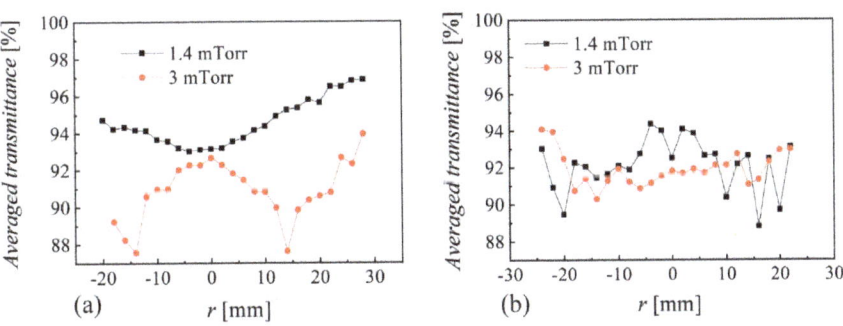

Figure 6. Spatial distribution of averaged transmittance for (**a**) $Z = 35$ mm and (**b**) $Z = 45$ mm where $P_{RF} = 20$ W.

Another important parameter is the band gap with a theoretical value of 3.37 eV for ZnO and expected increase with up to 0.5 eV by Al doping. The band gap of samples deposited at 1.4 and 3 mTorr and $Z = 35$ mm (20 W, 60 min) were calculated using Tauc's plot of transmittance spectra and are presented in Figure 7 with an evident correlation with the erosion tracks and also showing the highest values (above 3.4 eV) at the same locations with lowest resistivity values and highest mobility and carrier concentration.

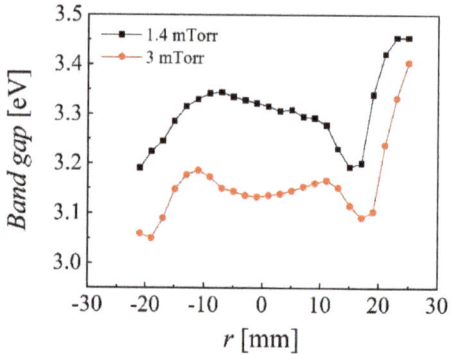

Figure 7. Spatial distribution of energy band gap for samples deposited at 1.4 and 3 mTorr where $Z = 35$ mm and $P_{RF} = 20$ W.

The spatial distribution of the (a) sheet resistance and (b) film thickens is presented in Figure 8 for different discharge powers at 1.4 mTorr, $Z = 35$ mm and 60 min deposition time. While the central part ($-10 < r < 10$ mm) was significantly affected by the torch-like discharge, noticed also in Figure 3, the lowest sheet resistance values (below 50 Ω/sq) were measured at the edge. The deep film thickness near $r = 0$ mm is caused by re-sputtering on the substrate, with no measurable values for powers above 20 W. The sheet resistance, Hall mobility, carrier concentration, resistivity and averaged transmittance for data points in Figure 8 at $r = 24$ mm are presented in Table 1 and show the lowest resistivity of 5.45×10^{-4} Ωcm, 17.3 cm²/Vs for mobility, 6.63×10^{20} cm^{-3} for carrier concentration and 88% for averaged transmittance. Such values are indeed very good for several applications that need TCO's with moderate properties, including low emissivity windows.

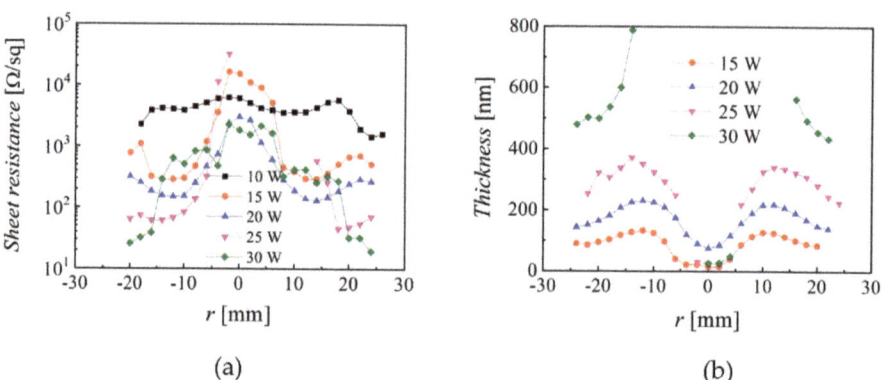

Figure 8. Spatial distribution for the (a) sheet resistance and (b) film thickness for different discharge powers (P_{RF}) where $p = 1.4$ mTorr, $Z = 35$ mm and a 60 min deposition time.

Table 1. Sheet resistance, Hall mobility, carrier concentration, resistivity and averaged transmittance for the data points in Figure 8 at r = 24 mm for different discharge powers.

RF Power [W]	Sheet Resistance/sq	Hall Mobility [cm^2/Vs]	Carrier Concentration [cm^{-3}]	Resistivity [cm]	Averaged Transmittance [%]
10	2320	2.62	5.07 × 10^{19}	4.69 × 10^{-2}	92.2
15	311	6.93	3.05 × 10^{20}	2.95 × 10^{-3}	89.6
20	175	7.15	3.84 × 10^{20}	2.27 × 10^{-3}	86.2
25	67.2	9.33	4.33 × 10^{20}	1.54 × 10^{-3}	87.7
30	13.6	17.3	6.63 × 10^{20}	5.45 × 10^{-4}	88.5

However, the challenge remains to attain such values all over the substrate. XRD and SEM performed on the 3 × 3 m^2 substrate pieces cut from the samples deposited at 1.4 (left) and 3 mTorr (right) presented in Figure 3a (20 W, Z = 35 mm, 60 min) are presented in Figure 9. Several crystalline structures can be identified with stronger peaks of Spinel (422) correlated with the erosion tracks and a well visible peak for Wurtzite (002) present only at the edge, corresponding to the lowest resistivity, as presented in Figure 4. Surface morphology by SEM reveals a higher roughness near the edge. Cross-SEM images have been presented elsewhere [20], including some showing a possible correlation with a transition in the Thornton diagram from zone 1 to zone 2 in a narrow pressure range.

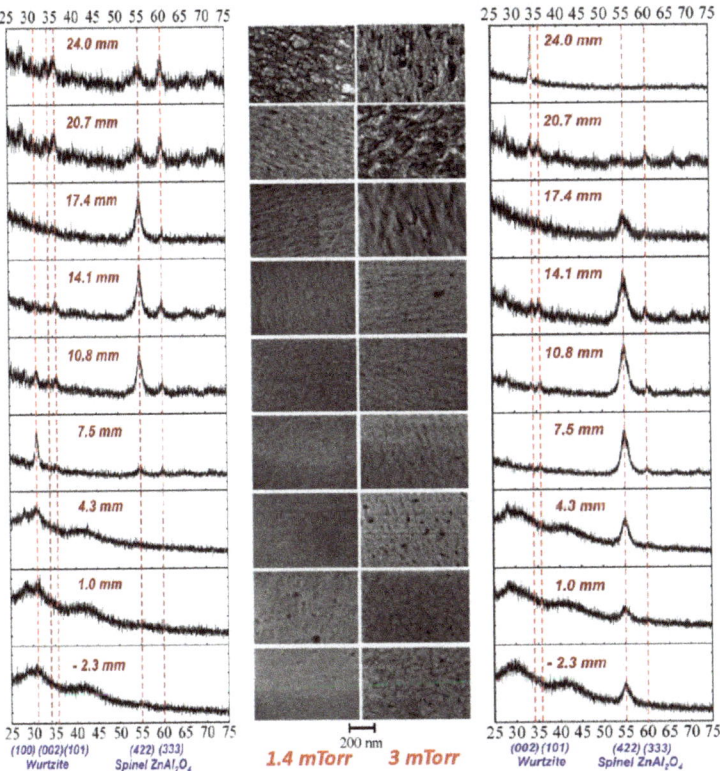

Figure 9. XRD and SEM performed on 3 × 3 m^2 substrate pieces cut from samples deposited at 1.4 (left) and 3 mTorr (right) presented in Figure 3a (20 W, Z = 35 mm, 60 min).

4. Discussion

AZO by magnetron sputtering has been under investigation for more than 30 years, with several detailed reports examining its non-uniformity aspects [45]. While the focus for an extended period was on reporting record values for the lowest resistivity, it became evident in recent years that resistivity values below 10^{-3} Ωcm are very difficult to obtain over an area comparable with that of the sputtering target for averaged transmittance values above 88%. Significant effort was also devoted to understanding the growth mechanism in correlation with the possible role of oxygen negative ions and atomic oxygen distribution [43,44]. The results presented in Section 3 reveal a very narrow range for deposition parameters where one can obtain reasonably good resistivity (10^{-3} Ωcm, with lower values outside the zone mirrored by the erosion tracks on the substrate): low pressure (2–4 mTorr), short target-to-substrate distance (30–45 mm) and low RF power (20–35 W for a 2 inch target). Increasing the pressure above 4 mTorr produced higher resistivity values. A similar trend was observed by increasing the target-to-substrate distance (see Figure 2). The short distance (Z) and low pressure also limits the discharge power to below 30 W (higher powers result in significant re-sputtering at the central part of the sample). The correlation of resistivity with the erosion track is obvious (see Figures 3 and 8) and it has been reported before. The XRD and SEM investigation (Figure 9) shows a Wurtzite (002) structure and larger grains only at the edge of the sample $|r| > 20$ mm while the Spinel (422) was dominant at locations facing the erosion tracks. Due to the glass substrate's contribution, the EDX investigation gave no relevant trends [20]. In a more detailed recent investigation, it was concluded that Al content was directly correlated with compressive stress, and the spatial inhomogeneity and pressure dependence could be related to particle bombardment [20]. An important aspect presented in this work is the very large variations one can get (almost two orders in magnitude for resistivity) within a very small shift in location (5–10 mm). This suggests that any attempts to understand the growth mechanism while rotating the substrate or neglecting the role of the investigation location with respect to the erosion track cannot lead to meaningful conclusions. The possibility of obtaining better resistivity values in certain locations far from the erosion track is also known and was intentionally used in several configurations [2]. However, this approach will be difficult to be implemented in large area coatings. As shown in Figure 1a, the present results were obtained by depositing a 50 mm long sample through a 60 mm opening in the shutter. Combined with the rather short target-to-substrate distance, one can see a possible shadowing effect at the sample's ends. The measurements performed without the shutter exhibited higher resistivity values all over the sample, a fact that suggests an additional positive role by placing a grounded electrode near the cathode [47]. The main point of this work is the optoelectronic characterization of the deposited films with a spatial resolution of 2–3 mm, which reveals the importance of trying to understand the growth mechanism by a careful examination of the entire spatial distribution of both the thin film and plasma parameters. While surface characterization techniques such as XPS, TOF-SIMS and XRD allows one to perform spatially resolved analytical investigations, this possibility alone cannot unveil the growth mechanism due to the need for coupling with spatially resolved plasma diagnostics. While electrostatic probes are subject to contamination and electromagnetic field distortions [48,49], optical diagnostics need complex 2D laser-induced fluorescence setups [50] to reveal the needed information.

5. Conclusions

Spatially resolved optoelectronic parameters of AZO thin films deposited by RF magnetron sputtering without intentional substrate heating were performed with the aim of narrowing down the process parameters that give the best film properties. For proper use in applications, both a low resistivity (below 3×10^{-3} Ωcm) and high transmittance (above 88%) should be attained on substrates comparable with target size. The strong correlation of film properties with the erosion tracks observed at a low pressure and a short target-to-substrate distance suggests that oxygen negative ions could be of higher relevance than the atomic oxygen concentration. A proper conclusion needs an adequate sputtering process design where one is able to control the negative ion energy.

Author Contributions: E.S. organized and performed the experiments and wrote the manuscript. All authors have read and agreed to the published version of the manuscript.

Funding: This work was partly supported by the SmartCoating project: 6151-00011B, financed by Innovation Fund Denmark.

Acknowledgments: Andrea Crovetto provided support for SEM investigation and Simone Sanna with XRD characterization.

Conflicts of Interest: The authors declare no conflict of interest.

References

1. Ginley, D.S.; Perkins, J.D. Transparent Conductors. In *Handbook of Transparent Conductors*; Ginley, D.S., Hosono, H., Paine, D.C., Eds.; Springer: New York, NY, USA, 2011; pp. 1–27.
2. Ellmer, K. Transparent conductive zinc oxide and its derivatives. In *Handbook of Transparent Conductors*; Ginley, D.S., Hosono, H., Paine, D.C., Eds.; Springer: New York, NY, USA, 2011; pp. 193–264.
3. Ellmer, K. Past achievements and future challenges in the development of optically transparent electrodes. *Nat. Photonics* **2012**, *6*, 809–817. [CrossRef]
4. Granqvist, C.G. Transparent conductors as solar energy materials: A panoramic review. *Sol. Energy Mater. Sol. Cells* **2012**, *91*, 1529–1598. [CrossRef]
5. Minami, T. Transparent conducting oxide semiconductors for transparent electrodes. *Semicond. Sci. Technol.* **2005**, *20*, S35–S44. [CrossRef]
6. Reinhard, P.; Chirila, A.; Blösch, P.; Pinezzi, F.; Nishiwaki, S.; Buecheler, S.; Tiwari, A.N. Review of progress toward 20% efficiency flexible CIGS solar cells and manufacturing issues of solar modules. *IEEE J. Photovolt.* **2013**, *3*, 572–580. [CrossRef]
7. Hajijafarassar, A.; Martinho, F.; Stulen, F.; Grini, S.; Lopez-Marrino, S.; Espindola-Rodriguez, M.; Dobeli, M.; Canulescu, S.; Stamate, E.; Gansukh, M.; et al. Monolitic thin-film chalcogenide-silicon tandem solar cells enabled by a diffusion barrier. *Sol. Energy Mater. Sol. Cells* **2019**, in press.
8. Agashe, C.; Kluth, O.; Schöpe, G.; Siekmann, H.; Hüpkes, J.; Rech, B. Optimization of the electrical properties of magnetron sputtered aluminum-doped zinc oxide films for opto-electronic applications. *Thin Solid Film.* **2003**, *442*, 167–172. [CrossRef]
9. Evcimen Duygulu, N.; Kodolbas, A.O.; Ekerim, A. Effects of argon pressure and r.f. power on magnetron sputtered aluminum doped ZnO thin films. *J. Cryst. Growth* **2014**, *394*, 116–125. [CrossRef]
10. Ghorannevis, Z.; Akbarnejad, E.; Salar Elahi, A.; Ghoranneviss, M. Application of RF magnetron sputtering for growth of AZO on glass substrate. *J. Cryst. Growth* **2016**, *447*, 62–66. [CrossRef]
11. Kim, D.K.; Kim, H.B. Room temperature deposition of Al-doped ZnO thin films on glass by RF magnetron sputtering under different Ar gas pressure. *J. Alloy. Comp.* **2011**, *509*, 421–425. [CrossRef]
12. Miao, D.; Jiang, S.; Zhao, H.; Shang, S.; Chen, Z. Characterization of AZO and Ag based films prepared by RF magnetron sputtering. *J. Alloy. Comp.* **2014**, *616*, 26–31. [CrossRef]
13. Patel, K.H.; Rawal, S.K. Influence of poFwer and temperature on properties of sputtered AZO films. *Thin Solid Film.* **2016**, *620*, 182–187. [CrossRef]
14. Prabhakar, T.; Dai, L.; Zhang, L.; Yang, R.; Li, L.; Guo, T.; Yan, Y. Effects of growth process on the optical and electrical properties in Al-doped ZnO thin films. *J. Appl. Phys.* **2014**, *115*, 083702. [CrossRef]
15. Shantheyanda, B.P.; Todi, V.O.; Sundaram, K.B.; Vijayakumar, A.; Oladeji, I. Compositional study of vacuum annealed Al doped ZnO thin films obtained by RF magnetron sputtering. *J. Vac. Sci. Technol. A* **2011**, *29*, 051514. [CrossRef]
16. Shih, N.F.; Chen, J.Z.; Jiang, Y.L. Properties and analysis of transparency conducting AZO films by using DC power and RF power simultaneous magnetron sputtering. *Adv. Mater. Sci. Eng.* **2013**, *2013*, 1–7. [CrossRef]
17. Sreedhar, A.; Kwon, J.H.; Yi, J.; Gwag, J.S. Improved physical properties of Al-doped ZnO thin films deposited by unbalanced RF magnetron sputtering. *Ceram. Int.* **2016**, *42*, 14456–14462. [CrossRef]
18. Sytchkova, A.; Luisa Grilli, M.; Rinaldi, A.; Vedraine, S.; Torchio, P.; Piegari, A.; Flory, F. Radio frequency sputtered Al:ZnO-Ag transparent conductor: A plasmonic nanostructure with enhanced optical and electrical properties. *J. Appl. Phys.* **2013**, *114*, 094509. [CrossRef]

19. Zubizarreta, C.; G.-Berasategui, E.; Ciarsolo, I.; Barriga, J.; Gaspar, D.; Martins, R.; Fortunato, E. The influence of target erosion grade in the optoelectronic properties of AZO coatings growth by magnetron sputtering. *Appl. Surf. Sci.* **2016**, *380*, 218–222. [CrossRef]
20. Crovetto, A.; Ottsen, T.S.; Stamate, E.; Kjær, D.; Schou, J.; Hansen, O. On performance limitations and property correlations of Al-doped ZnO deposited by radio-frequency sputtering. *J. Phys. D Appl. Phys.* **2016**, *49*, 295101. [CrossRef]
21. Cai, Y.; Liu, W.; He, Q.; Zhang, Y.; Yu, T.; Sun, Y. Influence of negative ion resputtering on Al-doped ZnO thin films prepared by mid-frequency magnetron sputtering. *Appl. Surf. Sci.* **2010**, *256*, 1694–1697. [CrossRef]
22. Guillén, C.; Herrero, J. High conductivity and transparent ZnO: Al films prepared at low temperature by DC and MF magnetron sputtering. *Thin Solid Film.* **2006**, *515*, 640–643. [CrossRef]
23. Hong, R.J.; Jiang, X.; Szyszka, B.; Sittinger, V.; Pflug, A. Studies on ZnO:Al thin films deposited by in-line reactive mid-frequency magnetron sputtering. *Appl. Surf. Sci.* **2003**, *207*, 341–350. [CrossRef]
24. Hong, R.; Xu, S. ZnO: Al films prepared by reactive mid-frequency magnetron sputtering with rotating cathode. *J. Mater. Sci. Technol.* **2010**, *26*, 872–877. [CrossRef]
25. Posadowski, W.M.; Wiatrowski, A.; Dora, J.; Radzimski, Z.J. Magnetron sputtering process control by medium-frequency power supply parameter. *Thin Solid Film.* **2008**, *516*, 4478–4482. [CrossRef]
26. Zhao, B.; Tang, L.D.; Wang, B.; Liu, B.W.; Feng, J.H. Optical and electrical characterization of gradient AZO thin film by magnetron sputtering. *J. Mater. Sci. Mater. Electron.* **2016**, *27*, 10320–10324. [CrossRef]
27. Chung, Y.M.; Moon, C.S.; Jung, M.J.; Han, J.G. The low temperature synthesis of Al doped ZnO films on glass and polymer using magnetron co-sputtering: Working pressure effect. *Surf. Coat. Technol.* **2005**, *200*, 936–939. [CrossRef]
28. Kumar, M.; Wen, L.; Sahu, B.B.; Han, J.G. Simultaneous enhancement of carrier mobility and concentration via tailoring of Al-chemical states in Al-ZnO thin films. *Appl. Phys. Lett.* **2015**, *106*, 241903. [CrossRef]
29. Linss, V. Comparison of the large-area reactive sputter processes of ZnO: Al and ITO using industrial size rotatable targets. *Surf. Coat. Technol.* **2016**, *290*, 43–57. [CrossRef]
30. Linss, V. Challenges in the industrial deposition of transparent conductive oxide materials by reactive magnetron sputtering from rotatable targets. *Thin Solid Film.* **2017**, *634*, 149–154. [CrossRef]
31. Nguyen, H.C.; Trinh, T.T.; Le, T.; Tran, C.V.; Tran, T.; Park, H.; Yi, J. The mechanisms of negative oxygen ion formation from Al-doped ZnO target and the improvements in electrical and optical properties of thin films using off-axis dc magnetron sputtering at low temperature. *Semicond. Sci. Technol.* **2011**, *26*, 105022. [CrossRef]
32. Oda, J.-I.; Nomoto, J.-I.; Miyata, T.; Minami, T. Improvements of spatial resistivity distribution in transparent conducting Al-doped ZnO thin films deposited by DC magnetron sputtering. *Thin Solid Film.* **2010**, *518*, 2984–2987. [CrossRef]
33. Rayerfrancis, A.; Bhargav, P.B.; Ahmed, N.; Bhattacharya, S.; Chandra, B.; Dhara, S. Sputtered AZO Thin Films for TCO and Back Reflector Applications in Improving the Efficiency of Thin Film a-Si:H Solar Cells. *Silicon* **2017**, *9*, 31–38. [CrossRef]
34. Sahu, B.B.; Han, J.G.; Hori, M.; Takeda, K. Langmuir probe and optical emission spectroscopy studies in magnetron sputtering plasmas for Al-doped ZnO film deposition. *J. Appl. Phys.* **2015**, *117*, 023301. [CrossRef]
35. Sato, Y.; Ishihara, K.; Oka, N.; Shigesato, Y. Spatial distribution of electrical properties for Al-doped ZnO films deposited by dc magnetron sputtering using various inert gases. *J. Vac. Sci. Technol. A* **2010**, *28*, 895–900. [CrossRef]
36. Wen, L.; Sahu, B.B.; Kim, H.R.; Han, J.G. Study on the electrical, optical, structural, and morphological properties of highly transparent and conductive AZO thin films prepared near room temperature. *Appl. Surf. Sci.* **2019**, *473*, 649–656. [CrossRef]
37. Xia, Y.; Wang, P.; Shi, S.; Zhang, M.; He, G.; Lv, J.; Sun, Z. Deposition and characterization of AZO thin films on flexible glass substrates using DC magnetron sputtering technique. *Ceram. Int.* **2017**, *43*, 4536–4544. [CrossRef]
38. Fumagalli, F.; Martí-Rujas, J.; Di Fonzo, F. Room temperature deposition of high figure of merit Al-doped zinc oxide by pulsed-direct current magnetron sputtering: Influence of energetic negative ion bombardment on film's optoelectronic properties. *Thin Solid Film.* **2014**, *569*, 44–51. [CrossRef]
39. Tiron, V.; Sirghi, L.; Popa, G. Control of aluminum doping of ZnO:Al thin films obtained by high-power impulse magnetron sputtering. *Thin Solid Film.* **2012**, *520*, 4305–4309. [CrossRef]

40. Mickan, M.; Helmersson, U.; Rinnert, H.; Ghanbaja, J.; Muller, D.; Horwat, D. Room temperature deposition of homogeneous, highly transparent and conductive Al-doped ZnO films by reactive high power impulse magnetron sputtering. *Sol. Energy Mater. Sol. Cells* **2016**, *157*, 742–749. [CrossRef]
41. Fu, S.W.; Chen, H.J.; Wu, H.T.; Hung, K.T.; Shih, C.F. Electrical and optical properties of Al:ZnO films prepared by ion-beam assisted sputtering. *Ceram. Int.* **2016**, *42*, 2626–2633. [CrossRef]
42. Pokorný, P.; Mišina, M.; Bulíř, J.; Lančok, J.; Fitl, P.; Musil, J.; Novotný, M. Investigation of the negative ions in Ar/O2 plasma of magnetron sputtering discharge with Al:Zn target by ion mass spectrometry. *Plasma Process. Polym.* **2011**, *8*, 459–464. [CrossRef]
43. Bikowski, A.; Welzel, T.; Ellmer, K. The correlation between the radial distribution of high-energetic ions and the structural as well as electrical properties of magnetron sputtered ZnO: Al films. *J. Appl. Phys.* **2013**, *114*, 223716. [CrossRef]
44. Bikowski, A.; Welzel, T.; Ellmer, K. The impact of negative oxygen ion bombardment on electronic and structural properties of magnetron sputtered ZnO: Al films. *Appl. Phys. Lett.* **2013**, *102*, 242106. [CrossRef]
45. Ellmer, K. Magnetron sputtering of transparent conductive zinc oxide: Relation between the sputtering parameters and the electronic properties. *J. Phys. D Appl. Phys.* **2000**, *4*, R17–R32. [CrossRef]
46. Agura, H.; Suzuki, A.; Matsushita, T.; Aoki, T.; Okuda, M. Low resistivity transparent conducting Al-doped ZnO films prepared by pulsed laser deposition. *Thin Solid Film.* **2003**, *445*, 263–267. [CrossRef]
47. Stamate, E. Method for Improving Optoelectronic Properties of Transparent Conductive Oxide Thin Films. Denmark Patent Application No. 70572, 18 September 2019.
48. Stamate, E.; Ohe, K. Influence of surface condition in Langmuir probe measurements. *J. Vac. Sci. Technol. A* **2002**, *20*, 661–666. [CrossRef]
49. Stamate, E. Status and challenges in electrical diagnostics of processing plasmas. *Surf. Coat. Technol.* **2014**, *260*, 401–410. [CrossRef]
50. Gao, J.; Nafarizal, N.; Sasaki, K. Measurement of Cu atom density in a magnetron plasma source using YBaCuO target by laser-induced fluorescence imaging spectroscopy. *J. Vac. Sci. Technol. A* **2006**, *24*, 2100–2104. [CrossRef]

© 2019 by the author. Licensee MDPI, Basel, Switzerland. This article is an open access article distributed under the terms and conditions of the Creative Commons Attribution (CC BY) license (http://creativecommons.org/licenses/by/4.0/).

Article

Antibacterial Nanostructured Ti Coatings by Magnetron Sputtering: From Laboratory Scales to Industrial Reactors

Rafael Alvarez [1,2,*], Sandra Muñoz-Piña [3], María U. González [4], Isabel Izquierdo-Barba [5,6], Iván Fernández-Martínez [3], Víctor Rico [1], Daniel Arcos [5,6], Aurelio García-Valenzuela [1], Alberto Palmero [1], María Vallet-Regi [5,6], Agustín R. González-Elipe [1] and José M. García-Martín [4,*]

1. Instituto de Ciencia de Materiales de Sevilla (CSIC-US), Américo Vespucio 49, 41092 Seville, Spain
2. Departamento de Física Aplicada I, Escuela Politécnica Superior, Universidad de Sevilla, c/Virgen de África 7, 41011 Seville, Spain
3. Nano4Energy SLNE, C/Jose Gutierrez Abascal 2, 28006 Madrid, Spain
4. Instituto de Micro y Nanotecnología, IMN-CNM, CSIC (CEI UAM + CSIC), Isaac Newton 8, 28760 Tres Cantos, Madrid, Spain
5. Dpto de Química en Ciencias Farmacéuticas, Facultad de Farmacia, Universidad Complutense de Madrid, Instituto de Investigación Sanitaria Hospital 12 de Octubre i+12, Plaza Ramón y Cajal s/n, 28040 Madrid, Spain
6. CIBER de Bioingeniería Biomateriales y Nanomedicina (CIBER-BBN), 28029 Madrid, Spain
* Correspondence: rafael.alvarez@icmse.csic.es (R.A.); josemiguel.garcia.martin@csic.es (J.M.G.-M.)

Received: 31 July 2019; Accepted: 27 August 2019; Published: 28 August 2019

Abstract: Based on an already tested laboratory procedure, a new magnetron sputtering methodology to simultaneously coat two-sides of large area implants (up to ~15 cm^2) with Ti nanocolumns in industrial reactors has been developed. By analyzing the required growth conditions in a laboratory setup, a new geometry and methodology have been proposed and tested in a semi-industrial scale reactor. A bone plate (DePuy Synthes) and a pseudo-rectangular bone plate extracted from a patient were coated following the new methodology, obtaining that their osteoblast proliferation efficiency and antibacterial functionality were equivalent to the coatings grown in the laboratory reactor on small areas. In particular, two kinds of experiments were performed: Analysis of bacterial adhesion and biofilm formation, and osteoblasts–bacteria competitive in vitro growth scenarios. In all these cases, the coatings show an opposite behavior toward osteoblast and bacterial proliferation, demonstrating that the proposed methodology represents a valid approach for industrial production and practical application of nanostructured titanium coatings.

Keywords: magnetron sputtering; oblique angle deposition; nanostructured titanium thin films; antibacterial coatings; osteoblast proliferation; industrial scale

1. Introduction

Addressing the problem of infection from the very first stage, i.e., inhibiting the formation of the bacterial biofilm, is a crucial step to prevent bone implant rejection. Recent studies indicate that nanostructured surfaces can be a less aggressive alternative to antibiotics to avoid infections [1,2], with the additional advantage of improving the behavior of osteoblasts, the cells that regenerate bone [3,4]. In this regard, the fabrication of nanostructured surfaces that may simultaneously favor the growth of osteoblasts and hinder bacterial proliferation represents a milestone in this research field with important implications, not only regarding the quality of life of patients but also by promoting a new generation of orthopedic implants. In the last few years, various alternatives have been proposed to induce such selective behavior, either by using nanostructures that incorporate drugs or bactericidal elements such as silver [5–7], or by surface processing with a strong corrugation at the nanoscale [8,9]. In our earlier work in 2015, we manufactured nanostructured coatings made of titanium (Ti) nanocolumns by oblique angle deposition (OAD) with magnetron sputtering onto the surface of Ti-6Al-4V discs, one of the alloys most commonly used in orthopedic implants. In vitro experiments showed that these nanocolumnar Ti coatings exhibited an efficient antibacterial behavior against *Staphylococcus aureus* (the bacterial adhesion decreased, and biofilm formation was prevented) without altering their biocompatibility (the osteoblasts proliferated and retained their mitochondrial activity) [10]. Moreover, in a more recent work, we have shown that these coatings also render similar antibacterial functionality against gram-negative bacteria (*Escherichia coli*) and, what is more important in order to have a direct impact in the field of medical implants, that these coatings could be prepared on small areas (~1 cm^2) either in a laboratory setup or in a semi-industrial scale equipment [11].

In this paper, we analyze the practical use of these coatings and their fabrication on larger scales, an aspect that is mandatory for the development of actual applications [12]. In general, regarding the minimization of costs and other economic and throughput issues, turning laboratory-size devices into operational market-ready products is a crucial engineering challenge that requires scaling up laboratory procedures to large area and mass production [13]. This issue is quite evident when using the magnetron sputtering (MS) method: By this technique, a plasma is made to interact with a solid target in a vacuum reactor, producing the sputtering of atomic species from a well-defined race-track region, and their deposition on a substrate located a few centimeters away [14]. In a classical MS configuration, the substrate is placed parallel to the target, producing the growth of highly compact and dense coatings, in a process that has been easily scaled up to mass-production methods by simply building larger versions of laboratory reactors [15]. Following this methodology, the magnetron sputtering technique has demonstrated being of great utility for the production of market ready devices in microelectronics [16], optical coatings [17], or sensors [18], among other devices and products [19–23]. Unlike the classical configuration, the OAD geometry promotes the arrival of sputtered atoms at the substrate along an oblique direction, inducing surface shadowing mechanisms and the formation of nanocolumnar arrays, which has been usually achieved by rotating the substrate with respect to the target in laboratory-scale procedures. However, due to the strongly non-linear nature of these atomistic processes, scaling up the OAD methodology from laboratory to mass-production scales is not straightforward, requiring the development of new approaches [24,25] and reactor designs [26], issues that have scarcely been addressed in the literature [27,28].

In this line, herein we develop a new engineering approach to coat with Ti nanocolumns two sides of bone plates with areas up to ~15 cm^2 that are commonly used to immobilize bone segments, and would be adequate for the development of this and other biomedical applications. To set up this new methodology we proceeded in the following way: We first analyzed the fundamental conditions leading to the formation of the nanocolumnar structures in a laboratory reactor, in particular, the energy and angular distribution of sputtered particles ejected from the magnetron target; then, based on these results, we proposed a new geometry to operate at oblique angles in semi-industrial reactors that reproduces these energy and momentum distributions at much larger scales. To prove the feasibility of the proposed design, we homogeneously and simultaneously coated the two sides of relatively large substrates and analyzed whether the antibacterial functionalities were the same as those obtained on surfaces manufactured in a laboratory MS reactor. In particular, two kinds of experiments were performed: Bacterial adhesion and biofilm formation, and osteoblasts–bacteria competitive in vitro assays, the latter also named the "Race for the Surface" competition [29].

2. Experimental Setup

The Ti coatings were first grown in a MS laboratory setup described in detail in reference [30] that from now forth will be dubbed l-reactor (see Figure 1a). It has a magnetron head (AJA Inc., MA, USA) with a circular 5 cm diameter Ti target and a cylindrical 9 cm long metallic chimney that collimates the flux of sputtered material and traps many of the thermalized atoms. The base pressure in the reactor is in the order of 10^{-7} Pa and the distance between target and substrate is 22 cm. The parameters used to fabricate the columnar coatings in this reactor with Ar as sputter gas [30] are: Pressure = 0.15 Pa, power (DC discharge) = 300 W, and tilt angle of the substrate with respect to the target α = 80°. The semi-industrial scale reactor, which will be called i-reactor hereafter, operates at the company Nano4Energy (see Figure 1b). The target is rectangular and much larger (20 × 7.5 cm^2) and, as a result of its balanced magnetic configuration, exhibits a racetrack with the shape of a rectangle (the long and short sides being 13.5 and 4.2 cm, respectively) with lines that are about 3 mm wide.

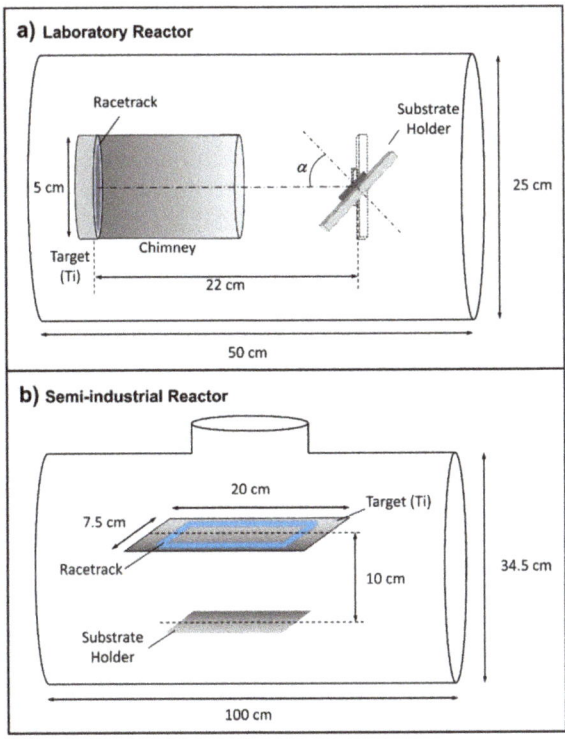

Figure 1. (a) Laboratory and (b) semi-industrial reactors employed to grow the Ti nanocolumns.

As a first step to scale up the growth conditions from the l-reactor to the i-reactor, we have employed as substrates fixation plates used in open trauma fractures that are known for their high postoperative infection rate (15% for patients with good health and more than 20% if they belong to risk groups). We coated two different fixation plates provided by Dr. Ricardo Larrainzar, Head of the Orthopedic Surgery and Traumatology Department at the "Infanta Leonor" University Hospital, Madrid. One of them was a new tubular plate from DePuy Synthes (made of stainless steel with length 5.2 cm, width 0.9 cm, and thickness 1 mm, with convex and concave sides), while the other was a pseudo-rectangular plate extracted from a patient and properly sterilized (with length 12 cm, width 1.3 cm, and thickness 4 mm). For depositions in the i-reactor, we used the following methodology: In the first stage, the plate was immersed in the plasma for cleaning purposes (pulsed DC voltage at 150 KHz, −500 V bias voltage, and a pressure of 1.2 Pa), after which the plate was left to cool down for 30 minutes. In the second stage, the Ti coating was deposited using the particular geometrical configuration presented in the Results and Discussion section. The deposition conditions were: Ar pressure = 0.4 Pa, power (DC discharge) = 325 W, and time = 25 min. Under these conditions, the deposition rate was ~12 nm/min and the thickness of the films about 300 nm. Finally, for the competitive studies between bacteria and osteoblasts, medical grade Ti-6Al-4V disks were also coated in the i-reactor and used for comparison with the large area sample results.

The microstructure of the coatings was studied with two different techniques: Scanning electron microscopy (SEM) with a Verios 460 field emission microscope (FEI Company, Hillsboro, OR, USA) using secondary electron detection, and atomic force microscopy (AFM) with a Dimension Icon microscope (Bruker Corporation, Billerica, MA, USA) that operates in a non-contact mode and type PPP-FM commercial probes,) (Nanosensors, Neuchâtel, Switzerland).

To check the antibacterial properties of the coatings, the DePuy Synthes bone plate was introduced in a solution of *S. aureus* bacterial strain (10^8 bacteria mL^{-1}) (15981 laboratory strain, ATCC, Manassas, VA, USA) and incubated for 24 h in a 66% tryptic soy broth (TSB) + 0.2% glucose environment to promote biofilm formation (20 g L^{-1} of Difco Bacto TBS (Becton Dickinson, Sparks, MD)). After 24 h, the plate was washed three times with sterile phosphate-buffered saline (PBS), stained with 3 μL of SYTO-9/propidium iodide mixture, incubated for 15 min and washed with PBS. To determine the formation of the biofilm, we used calcofluor, a fluorescent dye that has been used to stain the biofilm extracellular matrix. In this case, 1 mL of calcofluor solution (5 mg mL^{-1}) was used after the addition of the SYTO-9/propidium iodide mixture and was incubated for 15 min at room temperature. The formation of the biofilm was examined using a SP2 confocal laser scanning microscope (LEICA, Wetzlar, Germany). In this way, live and dead bacteria could be distinguished, with green and red, respectively, as well as the extracellular matrix of the biofilm with blue. Further details can be found in reference [10].

To further evaluate the antimicrobial activity of the nanostructured coatings, we carried out osteoblasts–bacteria competitive in vitro studies using the coated and uncoated regions of Ti-6Al-4V disks described above. For this purpose, co-cultures of MC3T3-E1 preosteoblast-like cells from mice (Sigma-Aldrich, San Luis, MO, USA) [31] and *S. aureus* [10] were co-cultured over uncoated surfaces and on surfaces coated with the Ti nanocolumns. Two different scenarios were simulated: i) Accidental infection (*S. aureus* concentrations of 10^2 cfu/mL), and ii) osteomyelitis scenario (*S. aureus* concentrations of 10^6 cfu/mL). In both cases, the *S. aureus* suspensions were mixed with 10^4 cells/mL of MC3T3-E1 preosteoblast, suspended in Todd Hewitt broth (THB) and complete Dulbecco's modified Eagle's medium (DMEM) and simultaneously seeded on the samples. After 6 h of culture, confocal microscopy studies were done and lactate dehydrogenase (LDH) levels were measured as a parameter of osteoblast destruction. In this regard, for confocal microscope, the actin of preosteoblast cytoskeleton was stained with Atto565-conjugated phalloidin (red) and both cell nuclei and bacteria were stained with DAPI (blue). Moreover, LDH level was determined in the culture medium, which is directly related to the rupture of the plasmatic membrane, (cell death) which, when broken, releases all organelles and enzymes present in the cytoplasm. Measurements were performed by using a commercial kit (Spinreact, Girona, Spain) having an absorbance at 340 nm with a UV–Visible spectrophotometer. Two measurements of three independent experiments were carried out. All data are expressed as means ± standard deviations of a representative of three independent experiments carried out in triplicate. Statistical analysis was performed using the Statistical Package for the Social Sciences (SPSS) version 19 software (IBM, Armonk, NY, USA). Statistical comparisons were made by analysis of variance (ANOVA). Scheffé test was used for post hoc evaluations of differences among groups. In all of the statistical evaluations, $p < 0.05$ was considered as statistically significant. The most representative confocal images are shown in this study.

3. Results and Discussion

3.1. From Laboratory to Industrial Reactors

In order to scale up the deposition procedure developed in the l-reactor we employed a well-known model to analyze the conditions required to grow the nanocolumnar films. In Figure 2a, we show the polar angle of incidence of sputtered Ti atoms on the tilted substrate in the l-reactor in our experimental conditions, as obtained by the SIMTRA code [32,33]. There, we can appreciate that the most probable angle of incidence over the substrate in this configuration is ~80°, which is above the calculated angular threshold of about ~70° required to promote the formation of the nanocolumns [30,34]. Moreover, as indicated in Figure 2a, there is a fraction of deposition species that arrives with lower angles of incidence, which corresponds to those atoms that have experienced collisions in the plasma and have altered their original steering [35]. The kinetic energy distribution function of these Ti atoms is shown in Figure 2b, and it is characterized by a long tail that extends up to energies above 10 eV, where the existence of numerous deposition species with kinetic energy above surface binding energy of Ti (~5 eV) is clear, i.e., with enough energy to induce kinetic energy-induced displacement processes of surface atoms upon deposition [30]. Using these distributions, we solved the model developed in reference [30] to account for the growth of Ti thin films by MS. The solution, presented in Figure 2c, shows a typical nanocolumnar array very similar to those experimentally obtained in reference [30], supporting that the necessary conditions for the growth of the Ti nanocolumns on a flat surface, as reported in reference [10], must also hold in the present case: i) The preferential angle of incidence of Ti sputtered species onto the surface must be centered at about ~80° with respect to the substrate normal, and ii) the kinetic energy distribution of deposition species must contain a significant fraction of atoms with energies above the binding energy of Ti surface atoms, i.e., ~5 eV.

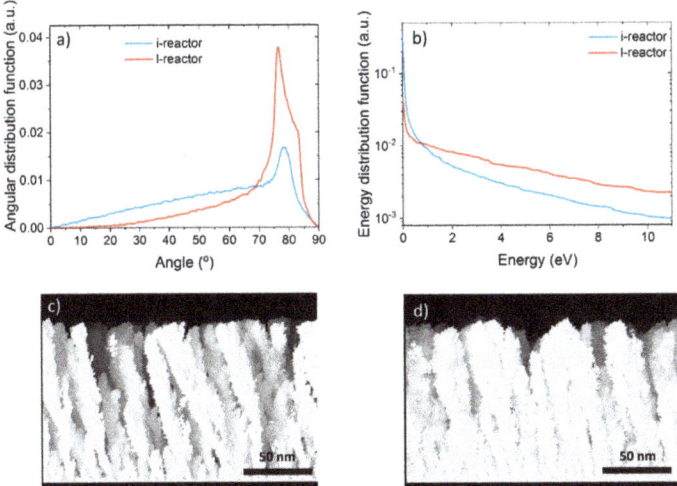

Figure 2. First row: (**a**) Polar angle distributions and (**b**) kinetic energy distributions of incident Ti atoms with respect to the surface normal in the l- and i-reactors (i.e., laboratory scale and semi-industrial scale, respectively). Second row: Solution of the model for the conditions in (**c**) the l-reactor and (**d**) the i-reactor.

Based on the results outlined above, we focused on reproducing both angular and kinetic energy distribution functions when operating the i-reactor on larger surfaces. In this way, and given the target and reactor geometry, we propose the geometrical arrangement shown in Figure 3. There, we placed the substrate perpendicular to the target, in such a way that atoms steaming from the racetrack may reach the substrate along an oblique angle of ~80°. Moreover, this particular configuration ensures that both sides of the substrate could be coated simultaneously. In Figure 2a, we show the calculated profile of the incident angle distribution of Ti species under this new configuration, where we can notice the similarities with that obtained in the l-reactor. This similarity extends to the kinetic energy distribution functions (see Figure 2b). In Figure 2d, we also show that the calculated nanostructure of the films in the i-reactor is formed by a nanocolumnar array, very similar to that obtained in the l-reactor (Figure 2c), suggesting the adequacy of the geometrical approach presented in Figure 3.

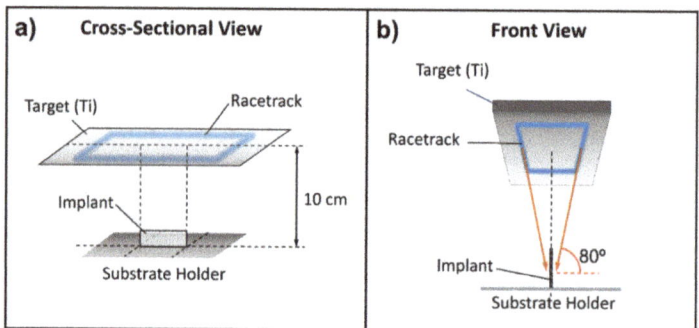

Figure 3. Proposed geometry ((a) cross-sectional and (b) front views) to coat the implants on two sides simultaneously with Ti nanocolumns.

3.2. Coating the Tubular Plate from DePuy Synthes

Following the geometrical configuration presented in Figure 3 and the conditions described in the Experimental Setup section, we coated the DePuy Synthes plate in the i-reactor. A mask protecting circa a quarter of the plate was employed to have an uncoated zone for the sake of comparison when performing in vitro analyses (see Figure 4). Scanning electron microscopy (SEM) and atomic force microscopy (AFM) were used to characterize the morphology of the coating, although the latter technique could only be applied on the convex side, as the tip holder of the microscope crashed with the lateral edges of the plate when approaching the concave surface. The uncoated zone presented a mirror-like brightness, indicative of small roughness. In agreement with this visual observation, both SEM and AFM images of this zone (not shown) indicate the absence of gaps or noticeable bumps on the surface, while the RMS roughness measured with the latter technique was 4 nm.

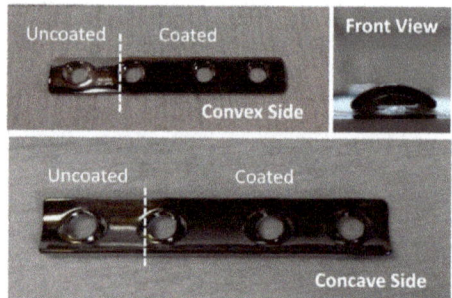

Figure 4. Different views of the DePuy Synthes plate coated in the i-reactor. A mask protecting about a quarter of the plate was used in order to have an uncoated zone to allow for comparison when performing in vitro analyses.

Figure 5a shows an AFM topographic map taken on the coated zone (convex side of the plate). It is worth noting the good homogeneity of the coating and the existence of a microstructure that consists of regularly separated nanocolumns. Figure 5b,c shows SEM images of the coating that were obtained on the convex and concave sides of the plate. The former shows a well-distributed and homogeneous Ti nanocolumnar array, very similar to those arrays obtained under laboratory conditions in references [10,30]. However, on the concave side, even though the coating is also homogeneous and consists of nanocolumns, these are now smaller both in diameter and length and are well packed, resembling a film with a rather compact structure. This means that, due to the curvature of the concave side of the substrate, sputtered atoms arrive at the surface with an angle of incidence below 80° at some locations, resulting in structures similar to those found in the l-reactor for lower angles of incidence [30]. This difference could be minimized by placing the substrate closer to the target in Figure 3b, thus promoting the arrival of sputtered species along higher polar angles of incidence.

Figure 5. Microscopy images of the DePuy Synthes plate after deposition of Ti nanocolumns: (**a**) Atomic force microscopy (AFM) topographic map of the convex side of the plate; (**b**) SEM image of the nanocolumnar structures in the convex and (**c**) concave side of the plate.

3.3. Coating of Pseudo-Rectangular Plate Extracted from a Patient

As an additional test of the geometrical arrangement presented in Figure 3, we analyzed the microstructure and morphology of a pseudo-rectangular plate extracted from a patient, as described in the Experimental Setup section. This plate was so large that it did not fit the entrance gate of the observation chamber of the SEM equipment and could only be analyzed by AFM. The initial gloss of the plate, which is rather matt instead of mirror-like, indicates that its roughness is high [36] (see Figure 6). To ascertain this, we performed a study of its morphology before the coating process. Figure 7a shows representative AFM images obtained in areas with different scale sizes (left and right of the figure, respectively) before deposition. On the higher magnification scale, the plate has a RMS roughness of 7 nm. However, in the image obtained in the same area but over a wider field of view (8 micron side), it can be seen that these flat areas are separated by deep cracks, with depths above one micron. This implies that, after the deposition process, most cracks will remain uncovered because their walls cast a shadowed region that avoids the arrival of most atoms inside, preventing the formation of nanocolumns. Consequently, the coatings will be inhomogeneous and there will be a large part of the implant surface (i.e., smooth areas of the initial surface) exhibiting well-formed nanocolumns, while a small percentage of it (deep cracks) will remain uncoated.

Following the sputtering process, the surface of the plate darkened considerably (c.f., middle and bottom panel in Figure 6), which indicates that a nanostructured coating has been successfully formed on both sides [37]. Figure 7b,c contains representative AFM images of the obtained coatings on both sides of the implant, upper and lower, respectively. They are composed of titanium nanocolumns, with a non-uniform distribution that depends on the morphology of the plate in each specific region: The columns grown on flat areas do have the same height, but those grown on the walls of the holes have lower height, as the initial surface was deeper. For example, Figure 7c shows an area with a very deep crack (depth about 1 micron) where it can be appreciated that the height of the columns is maximum at the top and gradually decreases when moving into the crack, until no columns are formed at the bottom. Overall, the columnar morphology of the coating is remarkably similar to that obtained on small substrates in the l-reactor in references [30] and [10].

Figure 6. Photographs of the pseudo-rectangular plate extracted from a patient, before and after deposition of Ti nanocolumns.

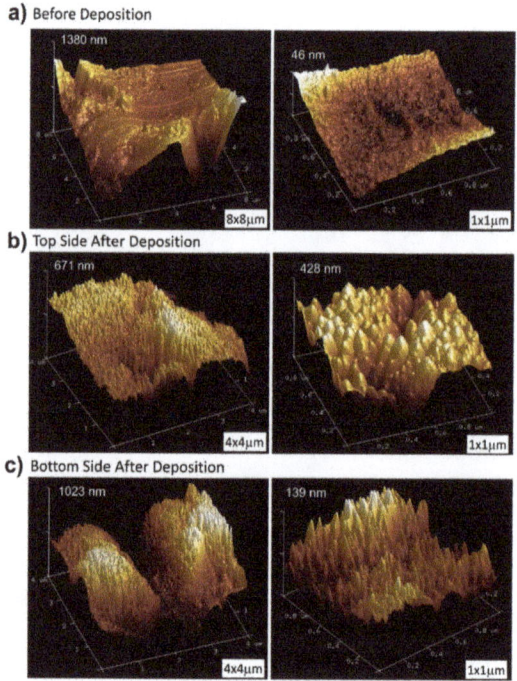

Figure 7. AFM images of the pseudo-rectangular plate extracted from a patient, obtained in areas with different size (left and right of the figure, respectively): (**a**) Before deposition; (**b**) top side after deposition; and (**c**) bottom side after deposition.

3.4. Bacterial Adhesion and Biofilm Formation

Once we had checked the nanocolumnar topography of the coatings produced in the i-reactor, we analyzed whether these maintain the same functionality as those produced in the l-reactor, i.e., if they are biocompatible and possess antibacterial capability. Following the bacterial growth procedure described in the Experimental Setup section, live and dead bacteria could be distinguished, with green and red, respectively, as well as the extracellular matrix of the biofilm with blue. Results appear in Figure 8, where we can clearly notice the bacterial proliferation on the uncoated region of the plate, which contains numerous living and dead bacteria, along with numerous blue staining, typical of extracellular matrix on the bacterial colonies. However, this blue stain does not appear in the coated zone in Figure 8, indicating the absence of bacterial biofilm in this case.

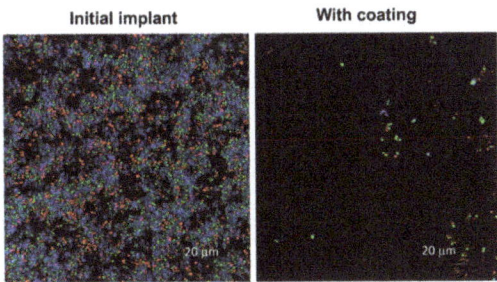

Figure 8. Antimicrobial activity in an osteoblasts–bacteria competitive in vitro scenario. Green corresponds to live bacteria, red to dead bacteria and blue corresponds to the extracellular matrix of the bacterial biofilm.

In order to further evaluate the antimicrobial activity of the nanostructured coatings, osteoblasts–bacteria competitive in vitro studies, already described in the Experimental Setup section, were also carried out in two different scenarios using the coated and uncoated regions of Ti-6Al-4V disks.

3.4.1. Accidental Infection Scenario

In this first case scenario, the MC3T3-E1/*S. aureus* ratio seeded was 100:1. Good osteoblast adhesion was observed in the uncoated and coated surfaces (Figure 9a,b). However, several lacunae could be observed in the case of the uncoated surface (Figure 9a), were colonies of *S. aureus* were present. On the contrary, the nanocolumnar surface appears almost fully coated by a MC3T3-E1 preosteoblast-like cells monolayer that reaches about 90% coverage, as can be seen in Figure 10a (right).

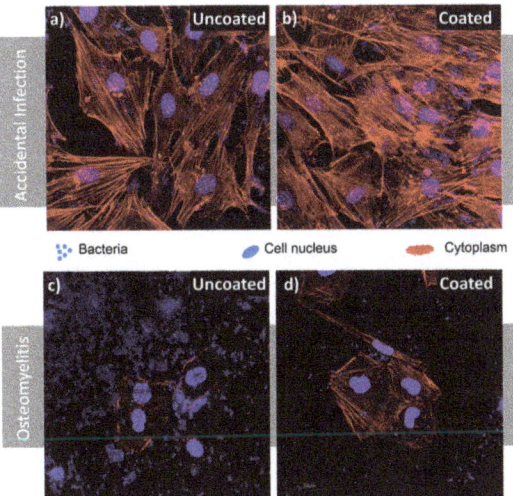

Figure 9. Competitive co-culture MC3T3-E1/*Staphylococcus aureus*: (**a**) 100:1 ratio (accidental infection scenario), uncoated region after 6 h; (**b**) 100:1 ratio (accidental infection scenario), coated region after 6 h; (**c**) ratio 1:100 (osteomyelitis scenario), uncoated region after 6 h; (**d**) ratio 1:100 (osteomyelitis scenario), coated region after 6 h.

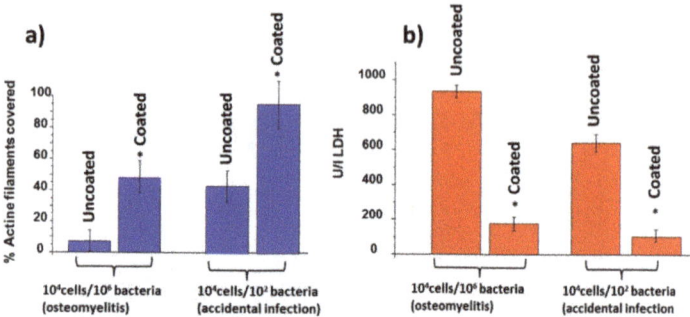

Figure 10. (a) Fraction of surface covered by preosteoblasts after 6 h under osteomyelitis (left) and accidental infection (right) scenarios; (b) lactate dehydrogenase (LDH) levels after 6 h under osteomyelitis (left) and accidental infection (right) scenarios.

LDH levels were measured as a parameter of cell destruction, illustrated in Figure 10b (right). There, it is evidenced that preosteoblast cell destruction is much higher on the uncoated surface than on the nanocolumnar coating under accidental infection scenarios.

3.4.2. Osteomyelitis Scenario

In this case, the MC3T3-E1/*S. aureus* ratio seeded was 1:100. After 6 h of culture, Ti-6Al-4V was covered by a significant amount of bacteria that had colonized most of the implant surface (Figure 9c,d). The number of osteoblast cells was significantly reduced and the cells exhibited rounded morphology with a low spreading degree. On the contrary, the nanocolumnar surface showed a higher degree of osteoblast proliferation and spread, thus occupying a significant amount of surface (around 50% as observed in Figure 10a, left). The very low presence of *S. aureus* in this sample, compared with Ti-6Al-4V must be highlighted. The LDH measurements also evidenced much higher preosteoblast destruction in the case of Ti-6Al-4V (see Figure 10b, left).

As a final comment, it is important to underline that the existence of the cracks on the fixation plates reported above implies that the coating is not fully homogeneous and, therefore, based on the results presented in [10], its efficiency as an antibacterial coating can be affected. This issue could be minimized by making use of a rather standard industrial technique, by which the substrate rotates around a certain axis to enhance the film homogeneity. In this manuscript, we have not attempted this approach, as we aim at scaling up an already reported laboratory technique that operates on static substrates. Nevertheless, it is likely that the existence of cracks is minimized when the substrate rotates around an axis parallel to the target (parallel to the substrate holder line in Figure 3b), so sputtered species may arrive at the film following a constant polar angle of incidence, but different azimuthal angles.

4. Conclusions

We developed a methodology based on a new geometry to coat two-sided surfaces with areas up to ~15 cm^2 with Ti nanocolumns by magnetron sputtering at oblique angles, and demonstrated its feasibility using a semi-industrial-scale reactor. This method was developed by calculating the necessary conditions for the growth of these structures in a laboratory-size reactor and reproducing them in a different geometry, suitable to coat larger areas in an industrial-scale reactor. These conditions were defined to control the incident angle distribution function of Ti atoms in the gaseous phase in such a way that they arrive at the surface along an oblique direction of about ~80–, and they possess a kinetic energy distribution function with a relevant proportion of deposition atoms with energies above the surface binding energy of Ti on the film surface.

After checking the homogeneity and features of the nanocolumnar structures deposited on different fixation plates on both sides, we analyzed the antibacterial functionality of the coating and demonstrated its equivalence to those produced in a laboratory reactor. In particular, two kinds of experiments were performed: Analysis of bacterial adhesion and biofilm formation, and osteoblasts–bacteria competitive in vitro scenarios, the latter also named "Race for the Surface" competition. In all these cases, we showed the opposite behavior of these surfaces toward osteoblast and bacterial proliferation and demonstrated that the proposed method represents a valid approach to coat large surfaces on both sides in industrial reactors, maintaining the same properties as laboratory-produced coatings on much smaller surfaces.

Author Contributions: Formal analysis, R.A., I.I.-B., and A.P.; funding acquisition, D.A., M.V.-R., A.R.G.-E., and J.M.G.-M.; investigation, R.A., S.M.-P., M.U.G., I.I.-B., I.F.-M., V.R., D.A., A.G.-V., A.P., and J.M.G.-M.; methodology, R.A., S.M.-P., M.U.G., I.I.-B., I.F.-M., V.R., D.A., A.G.-V., A.P., M.V.-R., A.R.G.-E., and J.M.G.-M.; project administration, J.M.G.-M.; resources, M.U.G., I.F.-M., V.R., A.P., and J.M.G.-M.; supervision, A.P. and J.M.G.-M.; visualization, R.A., S.M.-P., M.U.G., I.I.-B., A.P., and J.M.G.-M.; writing—original draft, R.A., I.I.-B., A.P., and J.M.G.-M.; writing—review and editing, R.A., S.M.-P., M.U.G., I.F.-M., D.A., A.P., M.V.-R., A.R.G.-E., and J.M.G.-M.

Funding: The authors acknowledge the financial support from the European Regional Development Funds program (EU-FEDER) and the MINECO-AEI (201560E055, MAT2014-59772-C2-1-P, MAT2016-75611-R, and MAT2016-79866-R and network MAT2015-69035-REDC). The authors also thank the University of Seville (V and VI PPIT-US) and Fundación Domingo Martínez for financial support. M.V.-R. also thanks the European Research Council (advanced grant VERDI; ERC-2015-AdG proposal 694160). The authors thank the MiNa Laboratory at IMN funded by CM (S2018/NMT-4291 TEC2SPACE), MINECO (CSIC13-4E-1794), and the EU (FEDER, FSE) for the support.

Acknowledgments: The authors acknowledge helpful discussions with Ricardo Larrainzar, Head of the Orthopedic Surgery and Traumatology Department at the "Infanta Leonor" University Hospital, Madrid. The National Center for Accelerators (Seville, Spain) is also acknowledged.

Conflicts of Interest: The authors declare no conflict of interest.

References

1. Puckett, S.D.; Taylor, E.; Raimondo, T.; Webster, T.J. The relationship between the nanostructure of titanium surfaces and bacterial attachment. *Biomaterials* **2010**, *31*, 706–713. [PubMed]
2. Jahed, Z.; Lin, P.; Seo, B.B.; Verma, M.S.; Gu, F.X.; Tsui, T.Y.; Mofrad, M.R. Responses of Staphylococcus aureus bacterial cells to nanocrystalline nickel nanostructures. *Biomaterials* **2014**, *35*, 4249–4254. [PubMed]
3. Webster, T.J.; Ejiofor, J.U. Increased osteoblast adhesion on nanophase metals: Ti, Ti6Al4V, and CoCrMo. *Biomaterials* **2004**, *25*, 4731–4739. [CrossRef] [PubMed]
4. De Oliveira, P.T.; Zalzal, S.F.; Beloti, M.M.; Rosa, A.L.; Nanci, A. Enhancement of in vitro osteogenesis on titanium by chemically produced nanotopography. *J. Biomed. Mater. Res. A* **2007**, *80*, 554–564. [CrossRef] [PubMed]
5. Kazemzadeh-Narbat, M.; Lai, B.F.; Ding, C.; Kizhakkedathu, J.N.; Hancock, R.E.; Wang, R. Multilayered coating on titanium for controlled release of antimicrobial peptides for the prevention of implant-associated infections. *Biomaterials* **2013**, *34*, 5969–5977. [CrossRef] [PubMed]
6. Mei, S.; Wang, H.; Wang, W.; Tong, L.; Pan, H.; Ruan, C.; Ma, Q.; Liu, M.; Yang, H.; Zhang, L.; et al. Antibacterial effects and biocompatibility of titanium surfaces with graded silver incorporation in titania nanotubes. *Biomaterials* **2014**, *35*, 4255–4265. [PubMed]
7. Cheng, H.; Xiong, W.; Fang, Z.; Guan, H.; Wu, W.; Li, Y.; Zhang, Y.; Alvarez, M.M.; Gao, B.; Huo, K.; et al. Strontium (Sr) and silver (Ag) loaded nanotubular structures with combined osteoinductive and antimicrobial activities. *Acta Biomater.* **2016**, *31*, 388–400. [CrossRef] [PubMed]
8. Bhadra, C.M.; Truong, V.K.; Pham, V.T.H.; Al Kobaisi, M.; Seniutinas, G.; Wang, J.Y.; Juodkazis, S.; Crawford, R.J.; Ivanova, E.P. Antibacterial titanium nano-patterned arrays inspired by dragonfly wings. *Sci. Rep.* **2015**, *5*, 16817. [CrossRef]
9. Bagherifard, S.; Hickey, D.J.; De Luca, A.C.; Malheiro, V.N.; Markaki, A.E.; Guagliano, M.; Webster, T.J. The influence of nanostructured features on bacterial adhesion and bone cell functions on severely shot peened 316L stainless steel. *Biomaterials* **2015**, *73*, 185–197. [CrossRef]

10. Izquierdo-Barba, I.; García-Martín, J.M.; Alvarez, R.; Palmero, A.; Esteban, J.; Pérez-Jorge, C.; Arcos, D.; Vallet-Regí, M. Nanocolumnar coatings with selective behavior towards osteoblast and Staphylococcus aureus proliferation. *Acta Biomater.* **2015**, *15*, 20–28. [CrossRef]
11. Cruz, D.M.; González, M.U.; Tien-Street, W.; Castro, M.F.; Crua, A.V.; Fernández-Martínez, I.; Martínez, L.; Huttel, Y.; Webster, T.J.; García-Martín, J.M.; et al. Synergic antibacterial coatings combining titanium nanocolumns and tellurium nanorods. *Nanomed. Nanotechnol. Boil. Med.* **2019**, *17*, 36–46.
12. Anders, A. Plasma and ion sources in large area coating: A review. *Surf. Coat. Technol.* **2005**, *200*, 1893–1906. [CrossRef]
13. Walter, C.; Sigumonrong, D.; El-Raghy, T.; Schneider, J. Towards large area deposition of Cr_2AlC on steel. *Thin Solid Films* **2006**, *515*, 389–393. [CrossRef]
14. Brauer, G.; Szyszka, B.; Vergöhl, M.; Bandorf, R. Magnetron sputtering—Milestones of 30 years. *Vacuum* **2010**, *84*, 1354–1359. [CrossRef]
15. Betz, U.; Olsson, M.K.; Marthy, J.; Escolá, M.; Atamny, F. Thin films engineering of indium tin oxide: Large area flat panel displays application. *Surf. Coat. Technol.* **2006**, *200*, 5751–5759. [CrossRef]
16. Choi, J.-H.; Kim, Y.-M.; Park, Y.-W.; Park, T.-H.; Jeong, J.-W.; Choi, H.-J.; Song, E.-H.; Lee, J.-W.; Kim, C.-H.; Ju, B.-K. Highly conformal SiO_2/Al_2O_3 nanolaminate gas-diffusion barriers for large-area flexible electronics applications. *Nanotechnology* **2010**, *21*, 475203. [CrossRef]
17. Oliva-Ramirez, M.; Gonzalez-Garcia, L.; Parra-Barranco, J.; Yubero, F.; Barranco, A.; Gonzalez-Elipe, A.R. Liquids Analysis with Optofluidic Bragg Microcavities. *ACS Appl. Mater. Interfaces* **2013**, *5*, 6743–6750. [CrossRef]
18. Martín, M.; Salazar, P.; Alvarez, R.; Palmero, A.; López-Santos, C.; González-Mora, J.; González-Elipe, A.R. Cholesterol biosensing with a polydopamine-modified nanostructured platinum electrode prepared by oblique angle physical vacuum deposition. *Sens. Actuators B Chem.* **2017**, *240*, 37–45. [CrossRef]
19. Ollitrault, J.; Martin, N.; Rauch, J.-Y.; Sanchez, J.-B.; Berger, F. Improvement of ozone detection with GLAD WO_3 films. *Mater. Lett.* **2015**, *155*, 1–3.
20. Sengstock, C.; Lopian, M.; Motemani, Y.; Borgmann, A.; Khare, C.; Buenconsejo, P.J.S.; Schildhauer, T.A.; Ludwig, A.; Köller, M. Structure-related antibacterial activity of a titanium nanostructured surface fabricated by glancing angle sputter deposition. *Nanotechnology* **2014**, *25*, 195101. [CrossRef]
21. Yoo, Y.J.; Lim, J.H.; Lee, G.J.; Jang, K.I.; Song, Y.M. Ultra-thin films with highly absorbent porous media fine-tunable for coloration and enhanced color purity. *Nanoscale* **2017**, *9*, 2986. [CrossRef] [PubMed]
22. Liu, Y.; Zhao, Y.; Feng, Y.; Shen, J.; Liang, X.; Huang, J.; Min, J.; Wang, L.; Shi, W. The influence of incident angle on physical properties of a novel back contact prepared by oblique angle deposition. *Appl. Surf. Sci.* **2016**, *363*, 252–258. [CrossRef]
23. Polat, B.; Keles, O. The effect of copper coating on nanocolumnar silicon anodes for lithium ion batteries. *Thin Solid Films* **2015**, *589*, 543–550. [CrossRef]
24. Godinho, V.; Moskovkin, P.; Alvarez, R.; Caballero-Hernández, J.; Schierholz, R.; Bera, B.; Demarche, J.; Palmero, A.; Fernández, A.; Lucas, S. On the formation of the porous structure in nanostructured a-Si coatings deposited by dc magnetron sputtering at oblique angles. *Nanotechnology* **2014**, *25*, 355705. [CrossRef] [PubMed]
25. Schierholz, R.; Lacroix, B.; Godinho, V.; Caballero-Hernández, J.; Duchamp, M.; Fernández, A. STEM–EELS analysis reveals stable high-density He in nanopores of amorphous silicon coatings deposited by magnetron sputtering. *Nanotechnology* **2015**, *26*, 75703. [CrossRef] [PubMed]
26. Garcia-Valenzuela, A.; Alvarez, R.; Rico, V.; Cotrino, J.; Gonzalez-Elipe, A.; Palmero, A. Growth of nanocolumnar porous TiO 2 thin films by magnetron sputtering using particle collimators. *Surf. Coat. Technol.* **2018**, *343*, 172–177. [CrossRef]
27. Krause, K.M.; Taschuk, M.T.; Brett, M.J. Glancing angle deposition on a roll: Towards high-throughput nanostructured thin films. *J. Vac. Sci. Technol. A* **2013**, *31*, 31507. [CrossRef]
28. Ziegler, N.; Sengstock, C.; Mai, V.; Schildhauer, T.A.; Köller, M.; Ludwig, A. Glancing-Angle Deposition of Nanostructures on an Implant Material Surface. *Nanomaterials* **2019**, *9*, 60. [CrossRef]
29. Pham, V.T.H.; Truong, V.K.; Orlowska, A.; Ghanaati, S.; Barbeck, M.; Booms, P.; Fulcher, A.J.; Bhadra, C.M.; Buividas, R.; Baulin, V.; et al. "Race for the Surface": Eukaryotic Cells Can Win. *ACS Appl. Mater. Interfaces* **2016**, *8*, 22025–22031. [CrossRef]

30. Alvarez, R.; Garcia-Martin, J.M.; Garcia-Valenzuela, A.; Macias-Montero, M.; Ferrer, F.J.; Santiso, J.; Rico, V.; Cotrino, J.; Gonzalez-Elipe, A.R.; Palmero, A. Nanostructured Ti thin films by magnetron sputtering at oblique angles. *J. Phys. D Appl. Phys.* **2016**, *49*, 045303. [CrossRef]
31. Izquierdo-Barba, I.; Santos-Ruiz, L.; Becerra, J.; Feito, M.J.; Fernández-Villa, D.; Serrano, M.C.; Díaz-Güemes, I.; Fernández-Tomé, B.; Enciso, S.; Sánchez-Margallo, F.M.; et al. Synergistic effect of Si-hydroxyapatite coating and VEGF adsorption on Ti6Al4V-ELI scaffolds for bone regeneration in an osteoporotic bone environment. *Acta Biomater.* **2019**, *83*, 456–466. [CrossRef] [PubMed]
32. Van Aeken, K. SIMTRA. Available online: www.draft.ugent.be (accessed on 29 July 2019).
33. Van Aeken, K.; Mahieu, S.; Depla, D. The metal flux from a rotating cylindrical magnetron: A Monte Carlo simulation. *J. Phys. D Appl. Phys.* **2008**, *41*, 205307. [CrossRef]
34. Dervaux, J.; Cormier, P.-A.; Moskovkin, P.; Douheret, O.; Konstantinidis, S.; Lazzaroni, R.; Lucas, S.; Snyders, R. Synthesis of nanostructured Ti thin films by combining glancing angle deposition and magnetron sputtering: A joint experimental and modeling study. *Thin Solid Films* **2017**, *636*, 644–657. [CrossRef]
35. Alvarez, R.; Romero-Gómez, P.; Gil-Rostra, J.; Cotrino, J.; Yubero, F.; Gonzalez-Elipe, A.; Palmero, A.; Romero-Gomez, P.; Gil-Rostra, J.; Gonzalez-Elipe, A.R. Growth of SiO$_2$ and TiO$_2$ thin films deposited by reactive magnetron sputtering and PECVD by the incorporation of non-directional deposition fluxes. *Phys. Status Solidi A* **2013**, *210*, 796–801. [CrossRef]
36. Stover, J.C. *Optical Scattering: Measurement and Analysis*; SPIE Optical Engineering Press: Bellingham, WA, USA, 1990. [CrossRef]
37. Vitrey, A.; Alvarez, R.; Palmero, A.; González, M.U.; García-Martín, J.M. Fabrication of black-gold coatings by glancing angle deposition with sputtering. *Beilstein J. Nanotechnol.* **2017**, *8*, 434–439. [CrossRef] [PubMed]

© 2019 by the authors. Licensee MDPI, Basel, Switzerland. This article is an open access article distributed under the terms and conditions of the Creative Commons Attribution (CC BY) license (http://creativecommons.org/licenses/by/4.0/).

Article

Design of Nanoscaled Surface Morphology of TiO$_2$–Ag$_2$O Composite Nanorods through Sputtering Decoration Process and Their Low-Concentration NO$_2$ Gas-Sensing Behaviors

Yuan-Chang Liang * and Yen-Chen Liu

Department of Optoelectronics and Materials Technology, National Taiwan Ocean University, Keelung 20224, Taiwan
* Correspondence: yuanvictory@gmail.com; Tel.: +886-24622192

Received: 1 August 2019; Accepted: 9 August 2019; Published: 11 August 2019

Abstract: TiO$_2$–Ag$_2$O composite nanorods with various Ag$_2$O configurations were synthesized by a two-step process, in which the core TiO$_2$ nanorods were prepared by the hydrothermal method and subsequently the Ag$_2$O crystals were deposited by sputtering deposition. Two types of the TiO$_2$–Ag$_2$O composite nanorods were fabricated; specifically, discrete Ag$_2$O particle-decorated TiO$_2$ composite nanorods and layered Ag$_2$O-encapsulated TiO$_2$ core–shell nanorods were designed by controlling the sputtering duration of the Ag$_2$O. The structural analysis revealed that the TiO$_2$–Ag$_2$O composite nanorods have high crystallinity. Moreover, precise control of the Ag$_2$O sputtering duration realized the dispersive decoration of the Ag$_2$O particles on the surfaces of the TiO$_2$ nanorods. By contrast, aggregation of the massive Ag$_2$O particles occurred with a prolonged Ag$_2$O sputtering duration; this engendered a layered coverage of the Ag$_2$O clusters on the surfaces of the TiO$_2$ nanorods. The TiO$_2$–Ag$_2$O composite nanorods with different Ag$_2$O coverage morphologies were used as chemoresistive sensors for the detection of trace amounts of NO$_2$ gas. The NO$_2$ gas-sensing performances of various TiO$_2$–Ag$_2$O composite nanorods were compared with that of pristine TiO$_2$ nanorods. The underlying mechanisms for the enhanced sensing performance were also discussed.

Keywords: sputtering; surface decoration; nanostructured surface; composite nanorods

1. Introduction

The development of chemosensors made from semiconductor oxides has recently become a key research topic [1,2]. Therefore, the development of highly responsive sensing oxide devices toward specific harmful gases has attracted interest in industry. For the gas sensor applications, one-dimensional (1D) metal oxides usually show better performance in comparison with their thin-film or bulk counterparts because of their high surface-to-volume ratio [3–6]. In particular, gas sensors based on 1D titanium dioxide (TiO$_2$) nanostructures have received considerable attention because they can be fabricated with diverse chemical and physical methods; moreover, TiO$_2$ has been shown to be favorable for the detection of diverse harmful gases and volatile organic vapors at elevated temperatures [5,7,8].

Recently, combining n-type oxides with p-type semiconductor oxides to form a heterogeneous structure has attracted great attention due to this combination's enhanced gas-sensing performance toward target gases [9–11]. The existence of the interfacial potential barrier at the p–n junctions of the heterogeneous structure play an important role in improving the gas-sensing performance of the constituent oxides. Several p–n junction-based sensors made from different material systems and configurations have been proposed. For examples, Woo et al. reported a discrete configuration of p-type Cr_2O_3 nanoparticles on the surfaces of ZnO nanowires; this p–n heterostructure enhances gas selectivity and sensitivity toward trimethylamine [11]. The decoration of NiO nanoparticles in porous SnO_2 nanorods remarkably enhances the gas-sensing response to ethanol as compared with pristine SnO_2, which can be attributed to the formation of NiO–SnO_2 p–n heterojunctions [12]. P-type Ag_2O phase-functionalized In_2O_3 nanowires shows an improved gas-sensing performance toward NO_2 gas [13]. However, reports on the incorporation of p-type oxides into n-type TiO_2 nanostructures to form a p–n junction gas sensor are limited in number.

In this study, 1D TiO_2–Ag_2O p–n heterogeneous structures are synthesized through the combination of hydrothermal growth and sputtering methods. Ag_2O is a p-type semiconductor oxide; it has previously been used as a gas-sensing material [13]. Moreover, Ag_2O crystals with various morphologies can be synthesized through chemical or physical routes [14–16]. Notably, Ag_2O particles are mostly synthesized through chemical routes but, using such chemical routes, it is hard to control the decoration morphology of the Ag_2O crystals on 1D nanostructures [15]. By contrast, the fabrication of Ag_2O through a physical method (sputtering) is advantageous concerning the control of the Ag_2O content, crystalline quality, and coverage morphology on 1D TiO_2. In this study, two types of TiO_2–Ag_2O composite nanorods are synthesized. By controlling the sputtering duration of the Ag_2O, discrete Ag_2O particle-decorated TiO_2 nanorods and Ag_2O layers encapsulating TiO_2 nanorods are fabricated. The Ag_2O coverage morphology effects on the low-concentration NO_2 gas-sensing performance of the TiO_2–Ag_2O p–n composite nanorods are systematically investigated in this study.

2. Materials and Methods

In this study, TiO_2 nanorods were grown on fluorine-doped SnO_2 (FTO) glass substrates. First, 0.25 mL of $TiCl_4$ and 19 mL HCl were added to 11 mL deionized water and then stirred to obtain a transparent solution for the hydrothermal growth of TiO_2 nanorods. The hydrothermal reaction was conducted at 180 °C for 3 h. For the preparation of TiO_2–Ag_2O composite nanorods, Ag_2O crystals were decorated onto the surfaces of the TiO_2 nanorod template via sputtering. Radiofrequency magnetron sputtering of Ag_2O was conducted using a silver metallic target in an Ar/O_2 (Ar:O_2 = 5:2) mixed ambient. The sputtering deposition temperature of the Ag_2O was maintained at 200 °C. The gas pressure during sputtering deposition was fixed at 20 mTorr and the sputtering power was fixed at 50 W for the silver target. Two sets of TiO_2–Ag_2O composite nanorods with Ag_2O sputtering durations of 130 s and 270 s were prepared; these were represented as TiO_2–Ag_2O-1 and TiO_2–Ag_2O-2, respectively, in this study. The sample configurations of the TiO_2–Ag_2O-1 and TiO_2–Ag_2O-2 composite nanorods are shown in Figure 1.

An X-ray diffractometer (XRD; D2 PHASER, Bruker, Karlsruhe, Germany) was used to analyze the crystal structures of the TiO_2–Ag_2O composite nanorods. Scanning electron microscopy (SEM; S-4800, Hitachi, Tokyo, Japan) and transmission electron microscopy (HRTEM; JEM-2100F, JEOL Tokyo, Japan) were used to characterize the morphology and detailed microstructures of the composite nanorod samples. The attached energy dispersive X-ray spectroscopy (EDS) of TEM was used to investigate the composition and composition distribution of the nanorod samples. Moreover, X-ray photoelectron spectroscopy (XPS; ULVAC-PHI XPS, ULVAC, Chigasaki, Japan) was used to characterize the elemental binding states of the synthesized samples.

The response of pure TiO_2 nanorods and the TiO_2–Ag_2O nanocomposites to NO_2 was tested in a vacuum test chamber. Silver electrodes were laid on the surfaces of the samples to form electric contacts for measurements. An Agilent B2911A meter measured the resistance variation of the nanorod sensors at a constant potential of 5 V as a function of time. Constant dry air was used as the carrier gas and the desired concentration of NO_2 gas (0.5, 1.5, 3.0 ppm) was introduced into the test chamber. A direct heating approach was used to operate the sensors at elevated temperatures in the range of 200–300 °C.

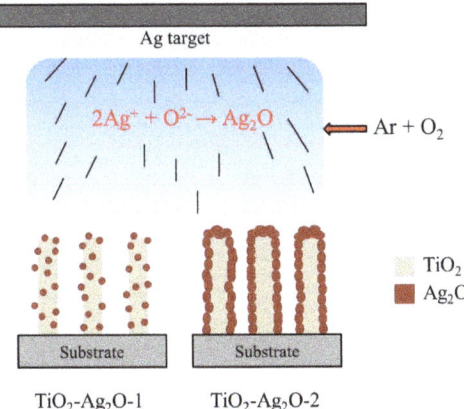

Figure 1. Sample configurations of the TiO$_2$–Ag$_2$O-1 and TiO$_2$–Ag$_2$O-2 composite nanorods synthesized with various sputtering durations of Ag$_2$O.

3. Results and Discussion

X-ray diffractometer (XRD) patterns of TiO$_2$–Ag$_2$O composite nanorods with various Ag$_2$O thin-film sputtering durations are shown in Figure 2. The distinct Bragg reflections centered at 27.46°, 36.05°, 41.22°, and 54.33° correspond to the crystallographic planes (110), (101), (111), and (211) of the rutile TiO$_2$ phase, respectively (JCPDS no. 00-004-0551). Moreover, the Bragg reflections centered at 32.72 ° and 37.98 ° are assigned to the crystallographic planes of cubic Ag$_2$O (111) and (200), respectively (JCPDS no. 00-012-0793). The XRD results reveal that highly crystalline rutile TiO$_2$-cubic Ag$_2$O composite nanorods were formed through the sputtering deposition of Ag$_2$O thin films onto the surfaces of the TiO$_2$ nanorods, and no other impurity peak was observed. As expected, the intensity of the Ag$_2$O Bragg reflections peaks increased with the increase of the Ag$_2$O thin-film sputtering duration, revealing an increased Ag$_2$O phase content in the composite nanorods.

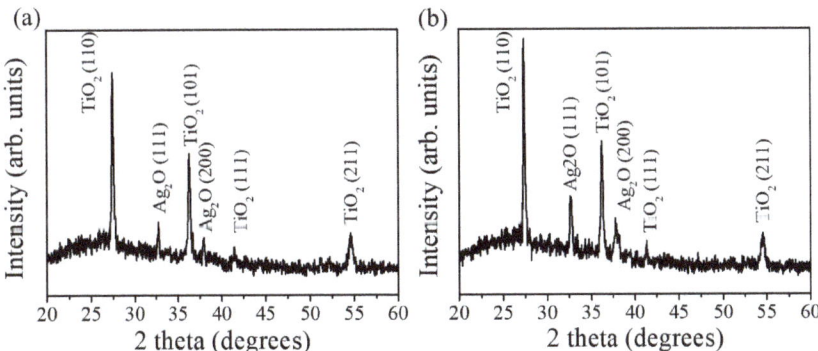

Figure 2. XRD patterns of various composite nanorods: (**a**) TiO$_2$–Ag$_2$O-1, (**b**) TiO$_2$–Ag$_2$O-2.

Figure 3a shows the scanning electron microscopy (SEM) image of the as-synthesized TiO_2 nanorods. The TiO_2 nanorods had a rectangular cross-section crystal feature with an average diameter of approximately 100 nm; the side facets of the nanorods were smooth. Figure 3b presents the SEM image of the TiO_2–Ag_2O-1 composite nanorods. The surface morphology of the composite nanorods reveals that a small amount of nanoparticle-like crystals was decorated onto the surfaces of the TiO_2 nanorods. The nanoparticle-like crystals dispersed separately on the surfaces of the TiO_2 nanorods. Figure 3c shows the SEM image of the TiO_2–Ag_2O-2 composite nanorods. The TiO_2 nanorods were homogeneously encapsulated by the aggregation of massive Ag_2O nanoparticles, resulting in the irregular-shaped cross-section crystal feature of the composite nanorods. Detailed TEM analyses were performed to further confirm the morphology change of the TiO_2–Ag_2O composite nanorods prepared at various Ag_2O sputtering durations.

Figure 3. SEM images of various nanorods: (a) TiO_2, (b) TiO_2–Ag_2O-1, (c) TiO_2–Ag_2O-2.

Figure 4a shows the low-magnification transmission electron microscopy (TEM) image of a TiO_2–Ag_2O-1 composite nanorod. A small amount of Ag_2O particles was dispersedly decorated on the surface of the TiO_2 nanorod via the sputtering growth of the Ag_2O. The high-resolution TEM images shown in Figure 4b,c indicate distinct lattice fringes in the Ag_2O particle. Moreover, the lattice fringe distance of approximately 0.24 nm was assigned to the crystallographic plane spacing of cubic Ag_2O (200). Figure 4d exhibits the selected area electron diffraction (SAED) pattern of several TiO_2–Ag_2O-1 composite nanorods. Several clear diffraction rings associated with (111) and (200) planes of the Ag_2O and (110), (111), and (211) planes of the rutile TiO_2 were observed in the SAED pattern. This demonstrates the good crystallinity of the composite nanorods and indicates that these composite nanorods have a polycrystalline nature. Figure 4e presents the EDS spectrum of a TiO_2–Ag_2O-1 nanorod. In addition to carbon and copper signals originating from the TEM grid, Ti, Ag, and O elements were detected in the selected heterostructure and no other impurity atom was detected. The EDS elemental mapping images taken from the TiO_2–Ag_2O-1 nanorod are presented in Figure 4f. The Ti signals were homogeneously distributed over the region of the nanorod template. By contrast, the Ag signals were mainly distributed on the outer region of the composite nanorod; the distribution of Ag signals was discrete and randomly decorated on the TiO_2 surface. A good Ag_2O particle-decorated TiO_2 nanorod with a dispersive particle decoration feature was obtained in the TiO_2–Ag_2O-1 composite nanorods.

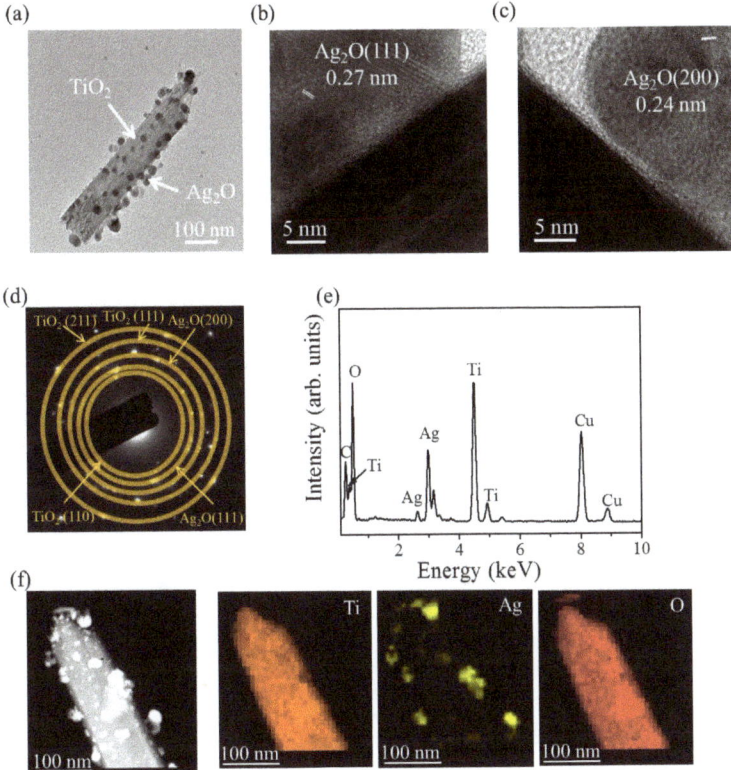

Figure 4. TEM analysis of the TiO$_2$–Ag$_2$O-1 composite nanorods: (**a**) Low-magnification TEM image of the TiO$_2$–Ag$_2$O-1 composite nanorod. (**b**,**c**) High-resolution TEM images taken from various regions of the composite nanorod. (**d**) Selected area electron diffraction (SAED) pattern of several TiO$_2$–Ag$_2$O-1 composite nanorods. (**e**) Energy dispersive X-ray spectroscopy (EDS) spectrum of the composite nanorod. (**f**) Ti, Ag, and O elemental mapping images taken from the selected composite nanorod.

Figure 5a shows the low-magnification TEM image of the TiO$_2$–Ag$_2$O-2 composite nanorod. In comparison with Figure 4a, the distribution density of the Ag$_2$O particles on the surface of the TiO$_2$ nanorod was denser and many particles were clustered, resulting in a rugged surface feature of the composite nanorod. The tiny Ag$_2$O particles aggregated together and fully encapsulated the surface of the TiO$_2$ nanorod. A clear heterointerface was observed between the TiO$_2$ and Ag$_2$O (Figure 5b,c). The distinct lattice fringes in the inner and outer regions of the composite nanorod in Figure 5b,c demonstrated a good crystallinity of the composite nanorod. The SAED pattern in Figure 5d supports the good crystallinity of the composite nanorods, as revealed in the high-resolution TEM images of the selected composite nanorod, and is also in agreement with the XRD result. Figure 5e shows the corresponding EDS spectrum of the TiO$_2$–Ag$_2$O-2 composite nanorod. Besides the carbon and copper signals and the elements of Ag, Ti, and O, no other impurity atom was detected from the selected composite nanorod. Notably, the relative intensity of the Ag signal was more intense than that of the TiO$_2$–Ag$_2$O-1 nanorod in Figure 4e, revealing a higher Ag content in TiO$_2$–Ag$_2$O-2 due to the prolonged sputtering duration of Ag$_2$O.

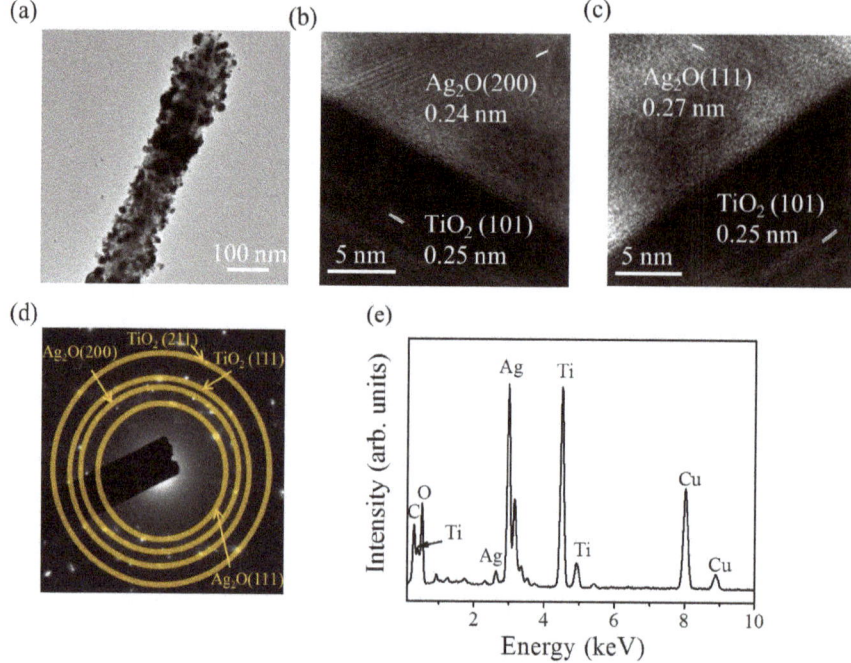

Figure 5. TEM analysis of the TiO$_2$–Ag$_2$O-2 composite nanorods: (**a**) Low-magnification TEM image of the TiO$_2$–Ag$_2$O-2 composite nanorod. (**b**,**c**) High-resolution TEM images taken from various regions of the composite nanorod. (**d**) SAED pattern of several TiO$_2$–Ag$_2$O-2 composite nanorods. (**e**) EDS spectrum of the composite nanorod.

Figure 6a shows the X-ray photoelectron spectroscopy (XPS) survey scan spectrum of the TiO$_2$–Ag$_2$O-1 composite nanorods. The primary features include the Ti, Ag, and O peaks that originated from the TiO$_2$–AgO composites. The trace carbon contamination on the surface of the nanorod sample originated from exposure to ambient air. No impurity atoms were detected in the nanorod sample. Figure 6b exhibits the Ag 3d core-level doublet spectrum originating from the Ag$_2$O decorated via sputtering; two distinct features centered at approximately 367.7 and 373.7 eV respectively correspond to the Ag 3d$_{5/2}$ and Ag 3d$_{3/2}$ binding energies. These binding energies are consistent with the Ag–O binding values reported for the Ag$_2$O phase [17]. This indicates that the silver exists in the Ag$^+$ valence state in the sputtered Ag$_2$O nanoparticles on the composite nanorods studied herein. Figure 6c displays the Ti 2p core-level doublet spectrum associated with the TiO$_2$ nanorod template. The distinct two features were deconvoluted into four subpeaks. The subpeaks centered at 458.7 and 464.3 eV correspond to Ti 2p$_{3/2}$ and Ti 2p$_{1/2}$ peaks of the Ti^{4+} valance state, respectively. By contrast, the subpeaks with a relatively weak intensity centered at 457.6 and 463.3 eV correspond to Ti 2p$_{3/2}$ for Ti 2p$_{1/2}$ peaks of the Ti^{3+} valence state [5]. The presence of the mixed Ti^{4+}/Ti^{3+} valance state indicates the possible presence of oxygen vacancies in the surfaces of the as-synthesized TiO$_2$ nanorods [5,8]. The O1s spectrum of the composite nanorods is shown in Figure 6d. The asymmetric O1s spectrum was deconvoluted into three subpeaks centered at 532.5, 531.2, and 529.2 eV. Notably, the subpeaks centered at 529.2 and 531.2 eV are ascribed to the lattice oxygen in Ag$_2$O and TiO$_2$, respectively [18,19]. Moreover, the external absorbed –OH groups or water molecules on the surfaces of the composite nanorods are reflected by a subpeak at approximately 532.5 eV [20].

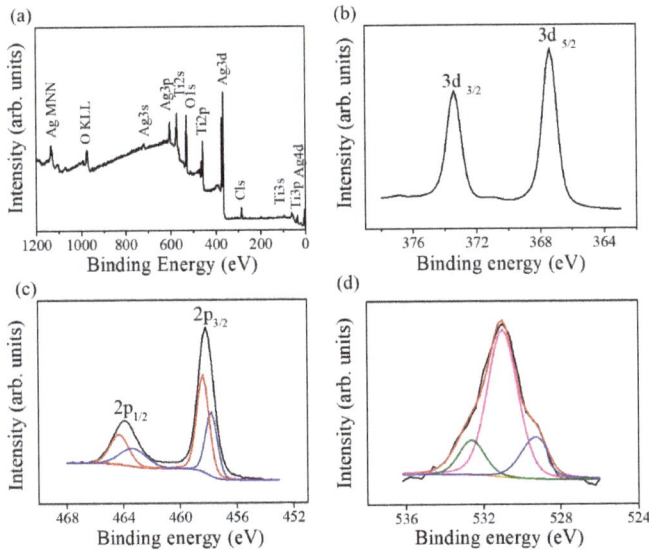

Figure 6. XPS analysis of the TiO$_2$–Ag$_2$O-1 composite nanorods: (**a**) Survey scan spectrum. (**b**) Ag 3d narrow scan spectrum. (**c**) Ti 2p narrow scan spectrum. The red curve is associated with the contribution of the Ti^{4+} valance state and the blue curve originated from the Ti^{3+} valence state. (**d**) O1s narrow scan spectrum. The blue and pink curves are ascribed to the lattice oxygen in Ag$_2$O and TiO$_2$, respectively. Moreover, the green curve is ascribed to external absorbed –OH groups or water molecules on the surfaces of the composite nanorods.

Figure 7 shows the temperature-dependent gas-sensing responses to NO$_2$ (1.5 ppm) of gas sensors made from TiO$_2$, TiO$_2$–Ag$_2$O-1, and TiO$_2$–Ag$_2$O-2 composite nanorods. For the NO$_2$ target gas, the n-type gas-sensing response of nanorod-based sensors is defined as Rg/Ra and the p-type gas-sensing response of nanorod-based sensors is defined as Ra/Rg, where Rg is the sensor resistance under target gas exposure and Ra is the sensor resistance with the removal of the target gas. The optimal operating temperature of oxide sensors to obtain the highest gas-sensing response is highly associated with the balance between the chemical reactions and the gas diffusion rate of the oxide surfaces [5]. The maximum responses of the TiO$_2$–Ag$_2$O-1 and TiO$_2$–Ag$_2$O-2 sensors to NO$_2$ were obtained at the operating temperature of 250 °C in this study. Meanwhile, a relatively high operating temperature of 275 °C was needed for the TiO$_2$ nanorods to obtain the maximum gas-sensing response under similar gas-sensing test conditions. Notably, the gas-sensing response versus operating temperature curve of TiO$_2$–Ag$_2$O-1 showed a distinct summit at 250 °C, differing substantially from the curves of the TiO$_2$ and TiO$_2$–Ag$_2$O-2 nanorod sensors. This result might be a sign of different gas-detecting mechanisms operating among the various nanorod-based sensors. Therefore, the optimal gas-sensing temperature of the fabricated composite nanorod sensors toward NO$_2$ was chosen as 250 °C in this study.

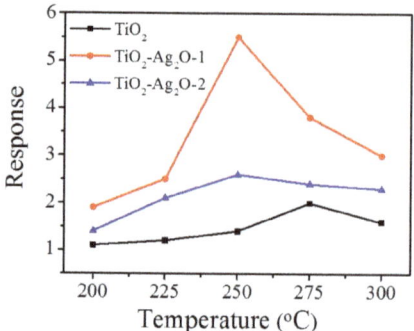

Figure 7. Temperature-dependent gas-sensing responses of various nanorod sensors exposed to 1.5 ppm NO_2.

Figure 8a–c shows the dynamic NO_2 gas-sensing response curves of the TiO_2, TiO_2–Ag_2O-1, and TiO_2–Ag_2O-2 sensors, respectively, upon exposure to 0.5–3.0 ppm NO_2. A sharp increase in sensor resistance was observed for the TiO_2 and TiO_2–Ag_2O-1 nanorod sensors upon exposure to NO_2; moreover, the sensor resistance decreased with the removal of the NO_2 target gas (Figure 8a,b). By contrast, the TiO_2–Ag_2O-2 showed an opposite sensor resistance variation upon exposure to NO_2 gas (Figure 8c). This indicates that the TiO_2 and TiO_2–Ag_2O-1 sensors showed an n-type conduction nature and the TiO_2–Ag_2O-2 sensor demonstrated a p-type conduction nature during the NO_2 gas-sensing tests. The aforementioned structural results reveal that the TiO_2–Ag_2O-1 sensor exhibited a morphology in which the Ag_2O particles were dispersedly distributed on the surfaces of the TiO_2 nanorods. The incomplete coverage of the Ag_2O particles on the TiO_2 surfaces of the TiO_2–Ag_2O-1 nanorods meant that, upon exposure to the NO_2 target gas, the n-type conduction dominated the material's gas-sensing behavior. By contrast, the TiO_2–Ag_2O-2 nanorods demonstrated a thick, full-coverage layer of Ag_2O clusters or aggregations on the surfaces of the TiO_2 nanorods. This morphology feature might account for the conduction and chemoresistive variation in the TiO_2–Ag_2O-2 sensor, which was dominated by p-type Ag_2O shell layers of the composite nanorods. A similar conduction type variation due to the p-type crystal coverage effect on the p–n heterogeneous oxides has been demonstrated in a ZnO–Cr_2O_3 system [11]. Comparatively, the TiO_2–Ag_2O-1 sensor exhibited the largest degree of sensor resistance variation before and after the introduction of the NO_2 gas under the given test conditions. Notably, the pristine TiO_2 sensor demonstrated the lowest sensor resistance variation size upon expose to NO_2 gas. The plot of NO_2 gas-sensing response versus NO_2 concentration for various TiO_2 nanorod-based sensors is shown in Figure 8d. The NO_2 gas-sensing response of the TiO_2–Ag_2O-1 sensor was approximately 3.1 upon exposure to 0.5 ppm NO_2. Moreover, the gas-sensing response of the TiO_2–Ag_2O-1 sensor increased to 7.6 upon exposure to 3.0 ppm NO_2. An approximate increase of the gas-sensing response by 2.4 times was observed with an increase in NO_2 concentration from 0.5 ppm to 3.0 ppm by the TiO_2–Ag_2O-1 sensor. By contrast, the TiO_2–Ag_2O-2 sensor exhibited gas-sensing responses of approximately 2.2 and 3.1 upon exposure to 0.5 ppm and 3.0 ppm NO_2, respectively; these response values are lower than those of the TiO_2–Ag_2O-1 sensor under similar test conditions. A concentration-dependent increment of the gas-sensing response for a low concentration range of 0.5–3.0 ppm NO_2 was less visible for the TiO_2–Ag_2O-2 sensor. Notably, the gas-sensing response of the pristine TiO_2 sensor at the same operating temperature did not show a response value larger than 2.0, revealing that the decoration of discrete or layered Ag_2O particles or aggregations on the surfaces of TiO_2 nanorods to form a p–n heterogeneous system is beneficial to the enhancement of the NO_2 gas-sensing response of TiO_2 nanorods. The gas-sensing response time of the nanorod-based sensors is defined as the duration required for an occurrence of a 90% change in sensor resistance upon exposure to the target gas, while the recovery time is the duration in which the sensor resistance drops by 90% from the maximal steady-state value, following the removal of the target gas.

The response times for the TiO$_2$, TiO$_2$–Ag$_2$O-1, and TiO$_2$–Ag$_2$O-2 nanorod sensors upon exposure to 0.5–3.0 ppm NO$_2$ gas ranged from 85 to 93 s. No substantial difference in the response times of various nanorod sensors exposed to different concentrations of NO$_2$ gas was observed. By contrast, a marked improvement in the recovery time of the TiO$_2$ nanorods sputtered with a coating of Ag$_2$O particles was visibly demonstrated. The recovery times of the pristine TiO$_2$ nanorod sensor ranged from 405 to 820 s after exposure to 0.5 to 3.0 ppm NO$_2$. Decreased recovery times were shown in the TiO$_2$–Ag$_2$O-2 nanorod sensor, which ranged from 191 to 280 s after exposure to 0.5 to 3.0 ppm NO$_2$. Notably, the TiO$_2$–Ag$_2$O-1 nanorod sensor exhibited a substantial decrease in the recovery time upon the removal of NO$_2$ gas; the recovery times ranged from 97 to 136 s in the NO$_2$ concentration range of 0.5 to 3.0 ppm. The size of the Ag$_2$O particles (or clusters) and their dispersibility are vital factors affecting gas-sensing performance, which lead to the highly effective desorption of surface-adsorbed ions with the removal of the target gas at elevated temperatures [21]. The TiO$_2$–Ag$_2$O-1 nanorod sensor exhibited the superior gas-sensing performance among the various nanorod sensors in this study. The cycling gas-sensing tests of the TiO$_2$–Ag$_2$O-1 nanorod sensor exposed to 1.5 ppm NO$_2$ at 250 °C are shown in Figure 8e. The result indicates that the TiO$_2$–Ag$_2$O-1 nanorod sensor had good reproducibility during multiple cycles of response and recovery. Figure 8f shows the across selectivity profiles of the TiO$_2$–Ag$_2$O-1 sensor upon exposure to 100 ppm H$_2$, 50 ppm C$_2$H$_5$OH, and 50 ppm NH$_3$ gases, as well as 3.0 ppm NO$_2$. The TiO$_2$–Ag$_2$O-1 sensor exhibited a highly selective gas-sensing response toward the low-concentration NO$_2$ gas as compared to the other various target gases.

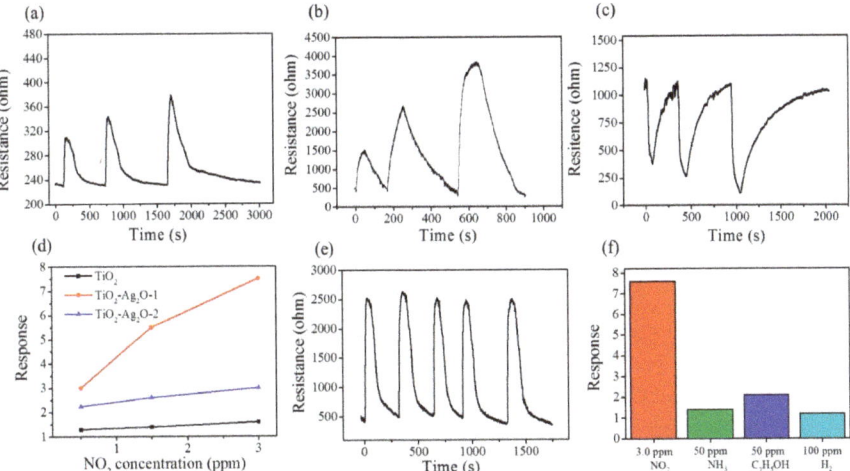

Figure 8. The dynamic response curves of various nanorod sensors to NO$_2$ gas ranging from 0.5 ppm to 3.0 ppm: (**a**) TiO$_2$, (**b**) TiO$_2$–Ag$_2$O-1, and (**c**) TiO$_2$–Ag$_2$O-2. (**d**) Summarized gas-sensing response values versus NO$_2$ concentration for various nanorod sensors. (**e**) Cycling gas-sensing tests of TiO$_2$–Ag$_2$O-1 upon exposure to 1.5 ppm NO$_2$ at 250 °C. (**f**) The across selectivity profiles of TiO$_2$–Ag$_2$O-1 upon exposure to various target gases.

The NO$_2$ gas-sensing performances of the sensors based on several TiO$_2$-based composite oxides are summarized in Table 1. Compared to previous works [22–25], the TiO$_2$–Ag$_2$O-1 nanorod sensor herein showed superior NO$_2$ gas-sensing performance under similar test conditions. The gas-sensing test results herein demonstrated that the TiO$_2$–Ag$_2$O composite nanorods decorated with discrete Ag$_2$O particles have potential for applications as NO$_2$ gas sensors at low concentrations. The possible surface chemisorption reactions occurring during the gas-sensing process of the TiO$_2$–Ag$_2$O composite nanorods upon exposure to NO$_2$ gas are described below:

$$NO_2 + e^- \rightarrow NO_2^-, \tag{1}$$

$$NO_2 + O_2^- + 2e^- \rightarrow NO_2^- + 2O^-. \tag{2}$$

Table 1. NO$_2$ gas-sensing performance of various TiO$_2$-based composites prepared using various methods in the operating temperature range of 200–300 °C [22–25].

Composite Nanorods	Synthesis Method	Operating Temperature (°C)	Concentration (ppm)	Response (Ra/Rg)	Detection Limit (ppm)	Response/Recovery Time (s)
TiO$_2$–Ag$_2$O (this work)	Hydrothermal and sputtering method	250	1.5	5.5	0.5	87/112
TiO$_2$–Er$_2$O$_3$	Sol-gel method	200	10	4.5	0.5	N/A
TiO$_2$–V$_2$O$_5$	Sol-gel and solvothermal method	200	2	0.8	N/A	N/A
TiO$_2$–MoO$_3$	Sol-gel method	300	2	2.3	0.5	120/180
TiO$_2$–Ga$_2$O$_3$	Sol-gel method	200	2	2.25	N/A	150/270

The NO$_2$ molecules capture electrons from the oxide surface to form NO$_2^-$ ions; this engenders the electron density variation of the oxides. By contrast, the surface-adsorbed NO$_2^-$ ions are desorbed with the removal of the NO$_2$ gas and, consequently, in this process the recovery of the initial conditions takes place. Notably, the contact of the TiO$_2$–Ag$_2$O oxides form p–n junctions at the hetero-interfacial regions. This additionally formed potential barrier in the TiO$_2$–Ag$_2$O composite nanorods explains the superior gas-sensing responses of the composite nanorods compared to that of the pristine TiO$_2$ nanorods. A similar formation of heterogeneous p–n junctions improves the gas-sensing responses of composite nanorods, as has been demonstrated in ZnO–ZnCr$_2$O$_4$, ZnO–Mn$_3$O$_4$, and ZnO–Cr$_2$O$_3$ p–n composite structures [9–11]. Furthermore, the reasons for the NO$_2$ gas-sensing response of the TiO$_2$–Ag$_2$O-1 sensor being higher than that of the TiO$_2$–Ag$_2$O-2 sensor at the given test conditions are explained by the schematic mechanisms exhibited in Figure 9. The schematic of the gas sensor device is also shown in Figure 9a. When the Ag$_2$O particles are coated on the surfaces of the TiO$_2$ nanorods in a discrete configuration, the randomly distribution of the depletion region at the interface of the p-Ag$_2$O and n-type TiO$_2$ will initially narrow the space of the conducting channel along the radial direction of the TiO$_2$ (Figure 9a). Moreover, the exposure of the free surfaces of the Ag$_2$O particles and TiO$_2$ rods in ambient air also initially lead to a surface hole accumulation layer and depletion layer, respectively. Furthermore, following the decoration of the Ag$_2$O particles in a continuous layer configuration on the surfaces of the TiO$_2$ nanorods, the conducting channel in the TiO$_2$ will also be narrowed (Figure 9b). After introducing NO$_2$ gas into the test chamber, the depletion region size at the TiO$_2$–Ag$_2$O hetero-interfacial region of the TiO$_2$–Ag$_2$O-1 sensor varies due to the surface-adsorbed NO$_2^-$ ions. Moreover, the surface depletion region of the TiO$_2$ nanorods is also thickened. The variation of the depletion size at different regions, further narrowing the conduction channel size of the TiO$_2$ nanorods, results in the increased sensor resistance of the TiO$_2$–Ag$_2$O-1 nanorod sensor. By contrast, the TiO$_2$–Ag$_2$O-2 nanorod sensor exhibits a p-type conduction gas-sensing behavior in this study. This reveals that the conduction channel size in the TiO$_2$ nanorod of the composite nanorod no longer plays a vital role affecting the chemoresistive variation upon exposure to NO$_2$ gas. It has also been shown that in core–shell ZnO–ZnMn$_2$O$_4$ and ZnO–Cr$_2$O$_3$ nanostructures, the p–n contact regions at the hetero-interfaces no longer play significant roles in the gas-sensing reaction [10,11]. The conduction path in the Ag$_2$O layer, by contrast, dominates the gas-sensing response of the TiO$_2$–Ag$_2$O-2 nanorod sensor. Notably, when the TiO$_2$–Ag$_2$O-2 nanorod is exposed to NO$_2$ gas, the accumulation layer in the Ag$_2$O layer thickens. This increases the carrier number in the p-type Ag$_2$O layer; therefore, a decreased sensor resistance is expected. However, the surface Ag$_2$O layer-dominated chemoresistive variation size of the TiO$_2$–Ag$_2$O-2 nanorod sensor is expected to be lower than that of the TiO$_2$–Ag$_2$O-1 nanorod sensor, which is dominated by the rugged conduction channel size in the TiO_2 core region upon exposure to NO_2 gas. Therefore, the superior NO_2 gas-sensing performance was obtained by the TiO_2–Ag_2O-1 nanorod sensor in this study.

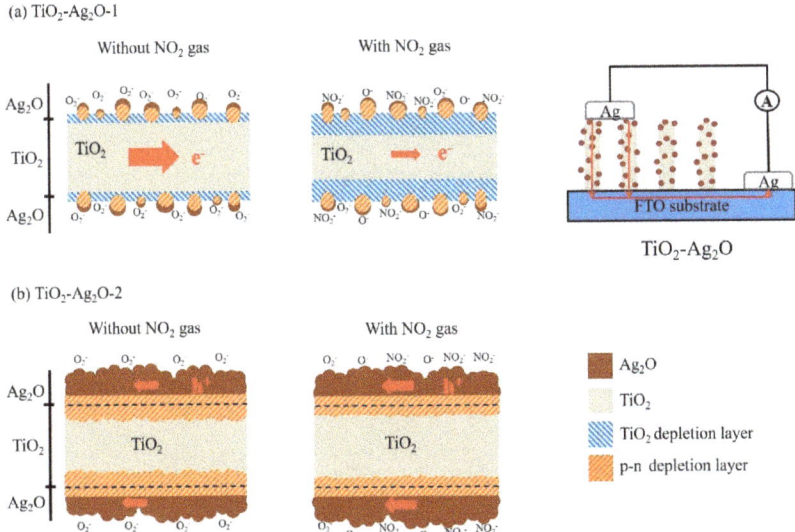

Figure 9. Schematic illustrations for possible gas-sensing mechanisms of (**a**) TiO_2–Ag_2O-1 and (**b**) TiO_2–Ag_2O-2 toward NO_2 gas.

4. Conclusions

In summary, TiO_2–Ag_2O composite nanorods were synthesized through the combination of hydrothermal growth and sputtering methods. The structural analysis reveals that the as-synthesized TiO_2–Ag_2O composite nanorods have a high crystallinity. The electron microscopy analysis results demonstrate that a shorter Ag_2O sputtering duration causes the formation of TiO_2–Ag_2O composite nanorods decorated with discrete Ag_2O particles. Meanwhile, a prolonged Ag_2O sputtering process engenders the aggregation of numerous Ag_2O particles, which form a layered configuration on the composite nanorods. The formation of p–n junctions in the composite nanorods enhances their NO_2 gas-sensing performance as compared to pristine TiO_2 nanorods. Moreover, different gas-sensing mechanisms of the TiO_2–Ag_2O nanorods with various Ag_2O coverage morphologies account for the superior NO_2 gas-sensing responses of the TiO_2–Ag_2O-1 sensor at a low concentration range in this study.

Author Contributions: Methodology, Y.-C.L. (Yen-Chen Liu); formal analysis, Y.-C.L. (Yen-Chen Liu); writing—original draft preparation, Y.-C.L. (Yuan-Chang Liang); supervision, Y.-C.L. (Yuan-Chang Liang).

Funding: This research was funded by the Ministry of Science and Technology of Taiwan. Grant No. MOST 105-2628-E-019-001-MY3 and MOST 108-2221-E-019-034-MY3.

Conflicts of Interest: The authors declare no conflict of interest.

References

1. Liang, Y.C.; Chao, Y. Crystal phase content-dependent functionality of dual phase SnO_2–WO_3 nanocomposite films via cosputtering crystal growth. *RSC Adv.* **2019**, *9*, 6482–6493. [CrossRef]
2. Liang, Y.C.; Lee, C.M.; Loa, Y.J. Reducing gas-sensing performance of Ce-doped SnO_2 thin films through a cosputtering method. *RSC Adv.* **2017**, *7*, 4724–4734. [CrossRef]

3. Liang, Y.C.; Chao, Y. Enhancement of Acetone Gas-Sensing Responses of Tapered WO_3 Nanorods through Sputtering Coating with a Thin SnO_2 Coverage Layer. *Nanomaterials* **2019**, *9*, 864. [CrossRef] [PubMed]
4. Liang, Y.C.; Chang, C.W. Improvement of Ethanol Gas-Sensing Responses of ZnO–WO_3 Composite Nanorods through Annealing Induced Local Phase Transformation. *Nanomaterials* **2019**, *9*, 669. [CrossRef] [PubMed]
5. Liang, Y.C.; Xu, N.C.; Wang, C.C.; Wei, D.H. Fabrication of Nanosized Island-Like CdO Crystallites-Decorated TiO_2 Rod Nanocomposites via a Combinational Methodology and Their Low-Concentration NO_2 Gas-Sensing Behavior. *Materials* **2017**, *10*, 778. [CrossRef] [PubMed]
6. Zhao, Y.P.; Li, Y.H.; Ren, X.P.; Gao, F.; Zhao, H. The Effect of Eu Doping on Microstructure, Morphology and Methanal-Sensing Performance of Highly Ordered SnO_2 Nanorods Array. *Nanomaterials* **2017**, *7*, 410. [CrossRef]
7. Liang, Y.C.; Xu, N.C. Synthesis of TiO_2–ZnS nanocomposites via sacrificial template sulfidation and their ethanol gas-sensing performance. *RSC Adv.* **2018**, *8*, 22437–22446. [CrossRef]
8. Liang, Y.C.; Wang, C.C. Hydrothermally derived zinc sulfide sphere-decorated titanium dioxide flower-like composites and their enhanced ethanol gas-sensing performance. *J. Alloys Compound.* **2018**, *730*, 333–341. [CrossRef]
9. Liang, Y.C.; Hsia, H.Y.; Cheng, Y.R.; Lee, C.M.; Liu, S.L.; Lina, T.Y.; Chung, C.C. Crystalline quality-dependent gas detection behaviors of zinc oxide–zinc chromite p–n heterostructures. *CrystEngComm* **2015**, *17*, 4190–4199. [CrossRef]
10. Na, C.W.; Park, S.Y.; Chung, J.H.; Lee, J.H. Transformation of ZnO Nanobelts into Single-Crystalline Mn_3O_4 Nanowires. *ACS Appl. Mater. Interfaces* **2012**, *4*, 6565–6572. [CrossRef]
11. Woo, H.S.; Na, C.W.; Kim, D.; Lee, J.H. Highly sensitive and selective trimethylamine sensor using one-dimensional ZnO–Cr_2O_3 hetero-nanostructures. *Nanotechnology* **2012**, *23*, 245501. [CrossRef]
12. Suna, G.; Chen, H.L.; Li, Y.W.; Chen, Z.H.; Zhang, S.S.; Ma, G.Z.; Jia, T.K.; Cao, J.L.; Bala, H.; Wang, X.D.; et al. Synthesis and improved gas sensing properties of NiO-decorated SnO_2 microflowers assembled with porous nanorods. *Sens. Actuators B* **2016**, *233*, 180–192. [CrossRef]
13. Kim, H.W.; Na, H.G.; Kwak, D.S.; Cho, H.Y.; Kwon, Y.J. Enhanced Gas Sensing Characteristics of Ag_2O-Functionalized Networked In_2O_3 Nanowires. *Jpn. J. Appl. Phys.* **2013**, *52*, 10MD01. [CrossRef]
14. Rajabi, A.; Ghazali, M.J.; Mahmoudi, E.; Baghdadi, A.H.; Mohammad, A.W.; Mustafah, N.M.; Ohnmar, H.; Naicker, S.A. Synthesis, Characterization, and Antibacterial Activity of Ag_2O-Loaded Polyethylene Terephthalate Fabric via Ultrasonic Method. *Nanomaterials* **2019**, *9*, 450. [CrossRef] [PubMed]
15. Ma, S.S.; Xue, J.J.; Zhou, Y.; Zhang, Z. Photochemical synthesis of ZnO-Ag_2O heterostructures with enhanced ultraviolet and visible photocatalytic activity. *J. Mater. Chem. A* **2014**, *2*, 7272–7280. [CrossRef]
16. Li, C.; Hsieh, J.H.; Cheng, J.C.; Huang, C.C. Optical and photoelectrochemical studies on Ag_2O/TiO_2 double-layer thin films. *Thin Solid Films* **2014**, *570*, 436–444. [CrossRef]
17. Sarkar, D.B.; Ghosh, C.K.; Mukherjee, S.; Chattopadhyay, K.K. Three Dimensional Ag_2O/TiO_2 Type-II (p–n) Nanoheterojunctions for Superior Photocatalytic Activity. *ACS Appl. Mater. Interfaces* **2013**, *5*, 331–337. [CrossRef]
18. Kaushik, V.K. XPS core level spectra and Auger parameters for some silver compounds. *J. Electron Spectrosc. Relat. Phenom.* **1991**, *56*, 273. [CrossRef]
19. Ingo, G.M.; Dirè, S.; Babonneau, F. XPS studies of SiO_2-TiO_2 powders prepared by sol-gel process. *Appl. Surf. Sci.* **1993**, *70*, 230–234. [CrossRef]
20. Wei, N.; Cui, H.Z.; Song, Q.; Zhang, L.; Song, X.J.; Wang, K.; Zhang, Y.F.; Li, J.; Wen, J.; Tian, J. Ag_2O nanoparticle/TiO_2 nanobelt heterostructures with remarkable photo-response and photocatalytic properties under UV, visible and near-infrared irradiation. *Appl. Catal. B* **2016**, *198*, 83–90. [CrossRef]
21. Yang, T.L.; Yang, Q.Y.; Xiao, Y.; Sun, P.; Wang, Z.; Gao, Y.; Ma, J.; Sun, Y.F.; Lu, G.Y. A pulse-driven sensor based on ordered mesoporous Ag_2O-SnO_2 with improved H_2S-sensing performance. *Sens. Actuators B* **2016**, *228*, 529–538. [CrossRef]
22. Mohammadi, M.R.; Fray, D.J. Development of nanocrystalline TiO_2–Er_2O_3 and TiO_2–Ta_2O_5 thin film gas sensors: Controlling the physical and sensing properties. *Sens. Actuators B* **2009**, *141*, 76–84. [CrossRef]
23. Epifani, M.R.; Comini, E.B.; Díaz, R.; Force, C.; Siciliano, P.T.; Faglia, G.D. TiO_2 colloidal nanocrystals surface modification by V_2O_5 species: Investigation by 47,49Ti MAS-NMR and H_2, CO and NO_2 sensing properties. *Appl. Surf. Sci.* **2015**, *351*, 1169–1173. [CrossRef]

24. Galatsis, K.; Li, Y.X.; Wlodarski, W.; Comini, E.; Faglia, G.; Sberveglieri, G. Semiconductor MoO_3–TiO_2 thin film gas sensors. *Sens. Actuators B* **2001**, *77*, 472–477. [CrossRef]
25. Mohammadi, M.R.; Fray, D.J. Semiconductor TiO_2–Ga_2O_3 thin film gas sensors derived from particulate sol–gel route. *Acta Mater.* **2007**, *55*, 4455–4466. [CrossRef]

© 2019 by the authors. Licensee MDPI, Basel, Switzerland. This article is an open access article distributed under the terms and conditions of the Creative Commons Attribution (CC BY) license (http://creativecommons.org/licenses/by/4.0/).

Article

Simultaneous Thermal Stability and Ultrahigh Sensitivity of Heterojunction SERS Substrates

Lingwei Ma [1], Jinke Wang [1], Hanchen Huang [2,*], Zhengjun Zhang [3,*], Xiaogang Li [1] and Yi Fan [4]

[1] Institute of Advanced Materials & Technology, University of Science and Technology Beijing, Beijing 100083, China; mlw1215@ustb.edu.cn (L.M.); wjkgege@126.com (J.W.); lixiaogang@ustb.edu.cn (X.L.)
[2] College of Engineering, University of North Texas, Denton, TX 76207, USA
[3] Key Laboratory of Advanced Materials, School of Materials Science and Engineering, Tsinghua University, Beijing 100084, China
[4] Jiangsu Key Laboratory for Premium Steel Material, Nanjing Iron and Steel Co., Ltd., Nanjing 210035, China; fanyi@njsteel.com.cn
* Correspondence: Hanchen.Huang@unt.edu (H.H.); zjzhang@tsinghua.edu.cn (Z.Z.)

Received: 5 May 2019; Accepted: 21 May 2019; Published: 31 May 2019

Abstract: This paper reports the design of Ag-Al_2O_3-Ag heterojunctions based on Ag nanorods (AgNRs) and their applications as thermally stable and ultrasensitive substrates of surface-enhanced Raman scattering (SERS). Specifically, an ultrathin Al_2O_3 capping layer of 10 nm on top of AgNRs serves to slow down the surface diffusion of Ag at high temperatures. Then, an additional Ag layer on top of the capping layer creates AgNRs-Al_2O_3-Ag heterojunctions, which lead to giant enhancement of electromagnetic fields within the Al_2O_3 gap regions that could boost the SERS enhancement. As a result of this design, the SERS substrates are thermally stable up to 200 °C, which has been increased by more than 100 °C compared with bare AgNRs, and their sensitivity is about 400% that of pure AgNRs. This easy yet effective capping approach offers a pathway to fabricate ultrasensitive, thermally stable and easily prepared SERS sensors, and to extend SERS applications for high-temperature detections, such as monitoring in situ the molecule reorientation process upon annealing. Such simultaneous achievement of thermal stability and SERS sensitivity represents a great advance in the design of SERS sensors and will inspire the fabrication of novel hetero-nanostructures.

Keywords: surface-enhanced Raman scattering (SERS); glancing angle deposition (GLAD); heterojunctions; SERS sensitivity; thermal stability

1. Introduction

Surface-enhanced Raman scattering (SERS) is the foundation of a powerful spectroscopic technique for rapid and non-destructive determination of chemical [1], environmental [2,3] and biological [4] analytes at trace levels, even at the level of a single molecule [5]. To achieve high sensitivity of SERS, the substrates are typically nanoscale noble metals, such as Ag, Au, and Cu, that have superior plasmonic efficiency [6,7]. While the nanoscale dimension gives rise to high sensitivity, it also leads to thermal instability because nanostructures coarsen easily at elevated temperatures [8,9]. For example, Ag nanorods (AgNRs) array and Ag colloids coarsen so much that they fuse at a temperature as low as 50 °C [10,11], thus limiting their practical SERS applications, such as monitoring in situ the thermal crystallization and the catalysis process.

Aiming to preserve the high sensitivity and yet to achieve thermal stability of SERS substrates, we have recently proposed and demonstrated the capping of AgNRs using high-melting temperature Al_2O_3 [12]. In contrast to oxide coating that also improves thermal stability [13–16], the oxide capping minimizes the reduction of exposed metallic surfaces that provide SERS sensitivity. A capping layer of 10 nm Al_2O_3 gives an optimal combination of thermal stability up to 200 °C and slight reduction of sensitivity by 25% [12]. The improved thermal stability opens the door for SERS applications in high-temperature environments. However, high-temperature SERS sensing often involves the detection of monolayer molecular adsorption and interface interactions [17–19], and therefore requires ultrahigh sensitivity. While preserving the improved thermal stability, is it possible to increase the sensitivity by 100% or more?

To achieve this goal, we propose the design of Ag-Al_2O_3-Ag heterojunctions based on AgNRs. As conceptually illustrated in Figure 1, we first deposit AgNRs and cap them with Al_2O_3 layers, and then deposit additional Ag of optimal thickness. The AgNRs-Al_2O_3-Ag heterojunctions give rise to giant electromagnetic (EM) enhancement [20], which in turn leads to ultrahigh SERS sensitivity [21,22]. In this paper, we demonstrate the feasibility of this proposal using the glancing angle deposition (GLAD) technique [10,12]. Our experiments show that the proposed design leads to thermal stability up to 200 °C and SERS sensitivity up to 400% that of pure AgNRs. The SERS intensity of different AgNRs-Al_2O_3-Ag substrates first elevates with the Ag capping thickness, reaches a maximum at the Ag thickness of 150 nm, and declines beyond the critical thickness. The Raman signal of methylene blue (MB) is clearly measurable on the Ag-10Al_2O_3-150Ag substrate, even at a low concentration of 1×10^{-10} M. The SERS enhancement factor is on the order of 10^8, which is comparable to the best reported results of AgNRs-based substrates [23–25], and the stable temperature of AgNRs-Al_2O_3-Ag increases by more than 100 °C. As a straightforward application of the heterojunctions, we exploit them to characterize the reorientation process of 4-tert-butylbenzylmercaptan (4-tBBM) molecules at elevated temperatures.

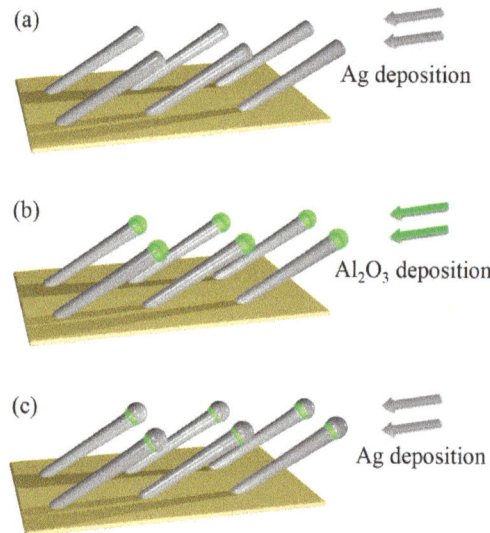

Figure 1. Schematic of (**a**) AgNRs deposition, (**b**) capping of AgNRs with Al_2O_3, (**c**) deposition of additional Ag.

2. Materials and Methods

2.1. Fabrication and Thermal Annealing of AgNRs-Al_2O_3-Ag Substrates

AgNRs are deposited on Si (001) substrates using the GLAD technique in an electron-beam system with a background vacuum level down to 10^{-5} Pa. During deposition, the incident angle of the vapor flux is set at 86° off the substrate normal. The nominal deposition rate is fixed at 0.75 nm/s. The nominal rate refers to the rate with zero-degree incidence angle, as read by a quartz crystal microbalance (QCM). The total nominal deposition is 1000 nm in thickness. After the deposition of AgNRs, without breaking the vacuum, the target is switched to Al_2O_3 in the deposition chamber. The incidence angle is set at 86° and the deposition rate is 0.1 nm/s. Our choice of Al_2O_3 thickness is based on the following two considerations. First, Al_2O_3 thickness needs to be sufficiently small to avoid the excessive coverage of Ag surfaces. Second, it needs to be large enough to maximize the thermal stability. The detailed optimization process of Al_2O_3 thickness is discussed in the Supporting Information (see Figures S1–S4). On the one hand, the SERS efficiency of AgNRs-Al_2O_3 substrates with 6 nm to 12 nm capping declined moderately with Al_2O_3 deposition. On the other hand, by depositing 10 nm or thicker Al_2O_3 onto AgNRs, the substrates exhibited no discernible morphology variation at 200 °C. Therefore, the total nominal deposition thickness of Al_2O_3 we used in this work is 10 nm [12].

Finally, the evaporation target is switched to Ag again in the chamber, the incident angle is kept at 86°, and the deposition rate is decreased to 0.3 nm/s for better coverage of the Al_2O_3. The total nominal deposition of Ag is set as 20, 50, 100, 125, 150, 175 and 200 nm, respectively, for each test. The AgNRs-Al_2O_3-Ag substrates are annealed on a hot plate at each given temperature for 15 min in air. The morphology evolution process during annealing is monitored in situ via the reflectivity variations, using the Optical Power Thermal Analyzer (OPA-1200, Wuhan Joule Yacht Science & Technology Co., Ltd., Wuhan, China).

2.2. Characterizations of AgNRs-Al_2O_3-Ag Substrates – Morphology, Structure, and SERS

The morphology and structure of the prepared SERS substrates are characterized using a scanning electron microscope (SEM, JEOL-JMS-7001F, Tokyo, Japan) and a high-resolution transmission electron microscope (HRTEM, JEOL-2011, Tokyo, Japan). SERS performance is characterized using an optical fiber micro-Raman system (i-Raman Plus, B&W TEK Inc., Newark, DE, United States), with MB and 4-tBBM as probing molecules. Raman spectra are collected based on an excitation laser of 785 nm in wavelength and of 100 mW in power, with a beam spot of ~80 microns in diameter. Before SERS characterizations, all substrates are immersed into analyte solutions for 30 min, washed thoroughly by deionized water to remove the residual molecules, and are then dried naturally in air. The data collection time for each spectrum is set to be five seconds and each SERS spectrum is obtained by measuring and averaging the signals collected from five different spots on a substrate.

2.3. FEM Simulation

Numerical simulations of AgNRs, AgNRs-Al_2O_3 and AgNRs-Al_2O_3-Ag were conducted using the finite element method (FEM) software COMSOL Multiphysics 5.2. Dimensional parameters were acquired from SEM observations, together with the excitation wavelength of 785 nm.

3. Results and Discussion

As the first set of results, we present the SEM and HRTEM images of AgNRs-Al_2O_3-Ag arrays of various Ag deposition thickness on top of AgNRs-Al_2O_3. As shown in Figure 2, the additional deposition of 50 nm Ag leads to the formation of Ag nanoparticles. As the deposition thickness increases to 100 nm, these nanoparticles merge to cover the Al_2O_3 capping. The coverage is more complete as the thickness increases to 150 nm. Further increase of the thickness to 200 nm results in the bridging of neighboring nanorods. Therefore, moderate thickness of deposition, around 150 nm, leads to the heterojunctions that we conceptually proposed in Figure 1.

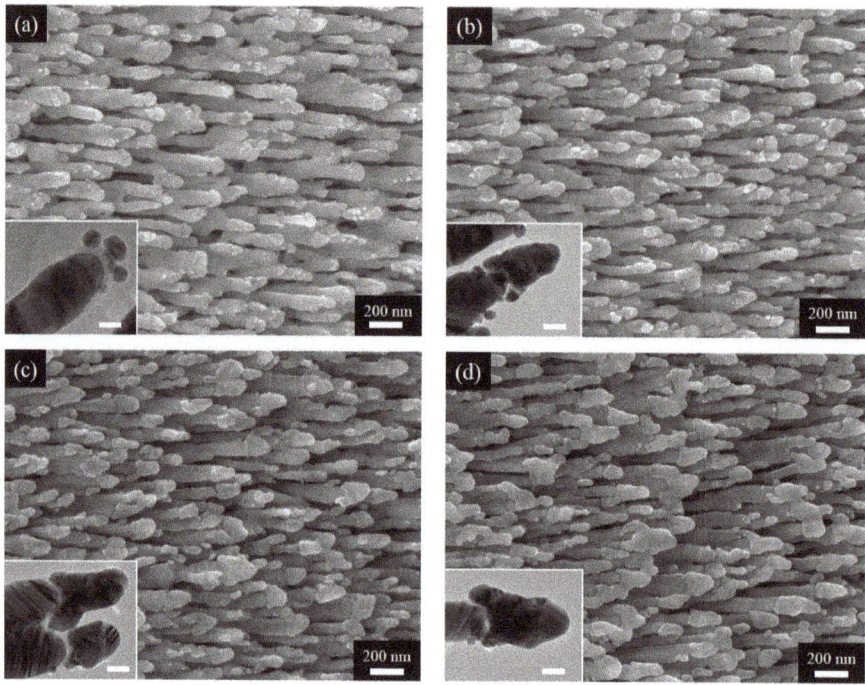

Figure 2. SEM images of AgNRs-Al$_2$O$_3$-Ag arrays with additional Ag deposition of (**a**) 50 nm, (**b**) 100 nm, (**c**) 150 nm, and (**d**) 200 nm; accompanying HRTEM images are included as insets with the scale bar being 20 nm.

Next, we examine the SERS sensitivity of AgNRs-Al$_2$O$_3$-Ag arrays. Figure 3a shows the SERS spectra of 1×10^{-6} M MB [26] on AgNRs-Al$_2$O$_3$-Ag arrays with various thickness of additional Ag deposition in the final stage. The spectrum on the AgNRs substrate is included for comparison. Characteristic Raman peaks are clearly seen in Figure 3a. In-plane bending of C–H is observed at 772 cm^{-1}, 886 cm^{-1}, and 1040 cm^{-1}, and in-plane ring mode of C–H is at 1300 cm^{-1}. 1181 cm^{-1} is assigned to the stretching of C–N. Besides, 1396 cm^{-1}, 1435 cm^{-1} and 1622 cm^{-1} correspond to the symmetric and asymmetric C-N stretches, as well as the C-C ring stretching, respectively. Indeed, the additional deposition of Ag increases the sensitivity, and an optimal amount is around 150 nm, consistent with the heterojunction morphology that is shown in Figure 2. Taking the characteristic peak at 1622 cm^{-1} as reference, Figure 3b shows the Raman intensity, normalized by that on bare AgNRs, as a function of the additional thickness of Ag deposition. The SERS intensity of different AgNRs-Al$_2$O$_3$-Ag substrates first elevates with the Ag capping thickness, reaches a maximum at the Ag thickness of 150 nm, and declines beyond the critical thickness. This sensitivity variation trend can be explained by different Ag morphologies: As Ag thickness increases, Ag layers grow from small particles into complete capping over nanorod tips, which could generate stronger EM field in the gap regions. When the deposition thickness exceeds 150 nm, Ag capping coalesces into large aggregations without clear tips and gaps, in which way the SERS activity declines to some extent. Remarkably, with the optimal Ag deposition thickness of 150 nm, the intensity (and thereby sensitivity) is about 400% that on bare AgNRs. For better comparison, we prepared two reference substrates, i.e., AgNRs coated with 150 nm Ag (AgNRs-150Ag) and 10 nm Al$_2$O$_3$ capped with 150 nm Ag (10Al$_2$O$_3$-150Ag). As shown in Figure S5, without Al$_2$O$_3$ gap, the AgNRs-150Ag film looks similar to bare AgNRs, with a little increase in nanorod length; the 10Al$_2$O$_3$-150Ag substrate produces many nanoparticles on Si surface instead of nanorods. The SERS intensities on AgNRs-Al$_2$O$_3$-Ag arrays are much higher than those obtained on AgNRs-150Ag and 10Al$_2$O$_3$-150Ag, as well as their intensity sum. These results manifest that without Al$_2$O$_3$ gap or underneath AgNRs, the nanocomposite could not generate very strong SERS enhancement. As for the AgNRs-Al$_2$O$_3$-Ag substrate, the intense interaction between AgNRs and Ag capping layers gives rise to a huge EM field within the Al$_2$O$_3$ gap regions, which is crucial for SERS enhancement. Figure 4 presents the FEM modeling results of (a) AgNRs; (b) AgNRs-10Al$_2$O$_3$; and (c) AgNRs-10Al$_2$O$_3$-150Ag, respectively. AgNRs possess intense EM enhancement at the tip and side of nanorods, while the EM field at the tip of AgNRs-10Al$_2$O$_3$ decreases slightly after Al$_2$O$_3$ capping. After further Ag deposition of 150 nm, there are evident "hot spots" at the gap between the AgNRs and the Ag capping layer, leading to stronger EM enhancement at the tip and side of this heterojunction structure. This simulation result suggests that the AgNRs-10Al$_2$O$_3$-150Ag array is a superior platform for maximizing the SERS performance.

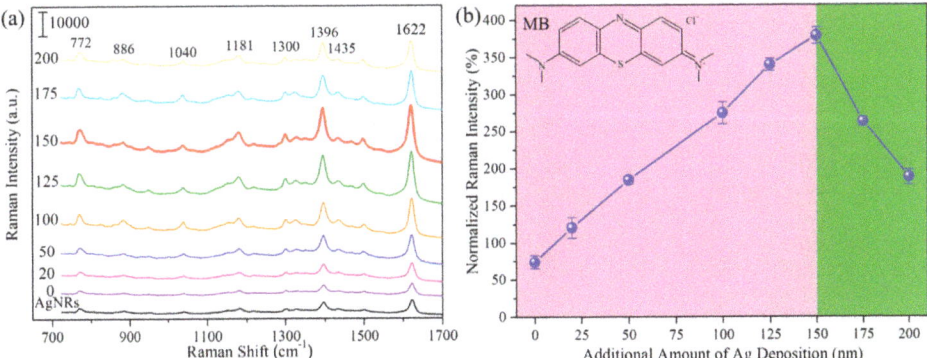

Figure 3. (a) SERS spectra of 1×10^{-6} M MB on bare AgNRs, and AgNRs-Al$_2$O$_3$-Ag arrays with additional Ag deposition of 0, 20, 50, 100, 125, 150, 175, and 200 nm; marked as AgNRs and the thickness of additional Ag deposition. (b) Raman intensity at 1622 cm^{-1}, normalized with respect to that on bare AgNRs, as a function of the additional thickness of Ag deposition.

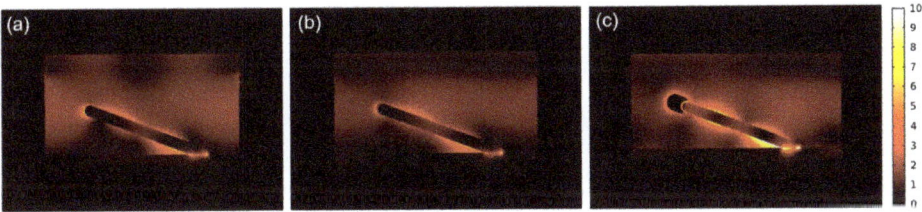

Figure 4. Localized electric field intensity distributions (indicated by the color bar) of (a) AgNRs, (b) AgNRs-10Al$_2$O$_3$; and (c) AgNRs-10Al$_2$O$_3$-150Ag, respectively.

After discussing the optimization of the proposed substrates, we now examine the SERS efficiency and quantification capacity of the designed heterojunctions. Taking AgNRs-Al$_2$O$_3$-Ag arrays with 150 nm additional Ag as the substrate, Figure 5a shows the SERS spectra of MB molecules of various concentrations, and Figure 5b shows the Raman peak intensity at 1622 cm^{-1} as a function of the concentration from 1×10^{-10} M to 1×10^{-6} M. Even at the concentration of 1×10^{-10} M, the Raman signal is clearly measurable and follows the linear dependence on concentration. The peak intensity increases linearly with concentration from 1×10^{-10} M to 1×10^{-8} M. The limit of detection (LOD) of MB is about 9.3×10^{-11} M based on a signal-to-background ratio of S/N = 3. At higher concentrations, the intensity increases slowly due to the saturated adsorption. As described in the Supporting Information, the SERS enhancement factor reaches 1.3×10^8, which is comparable to the best reported values of AgNRs-based substrates [23–25], reflecting the ultrahigh SERS efficiency of the AgNRs-Al$_2$O$_3$-Ag heterojunctions.

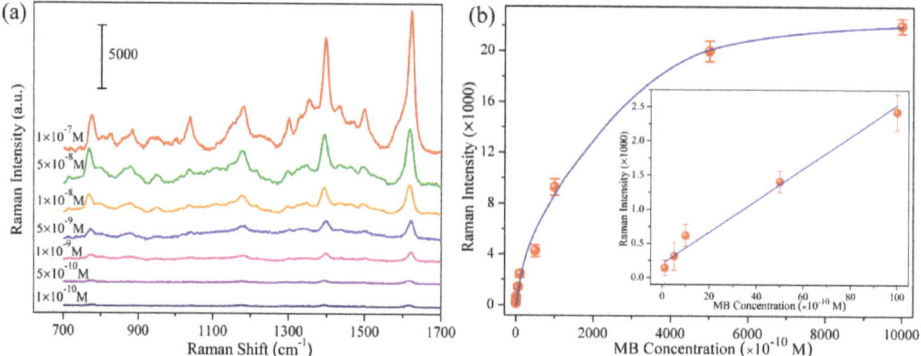

Figure 5. (a) SERS spectra of MB molecules on AgNRs-Al$_2$O$_3$-Ag substrates with 150 nm additional Ag; the thickness of MB molecules is marked by each spectrum; (b) Raman peak intensity at 1622 cm^{-1} as a function of MB concentrations, inserted: 1622 cm^{-1} peak intensity from 1×10^{-10} M to 1×10^{-8} M.

Having established the high sensitivity of AgNRs-Al$_2$O$_3$-Ag arrays with 150 nm additional Ag, we now examine their thermal stability upon annealing. Figure 6 shows the reflectivity of the AgNRs-Al$_2$O$_3$-Ag arrays as a function of the annealing temperature [16]. As expected, the Al$_2$O$_3$ capping layer still serves the purpose of slowing down the surface diffusion so as to increase the thermal stability up to 200 °C. In comparison with the morphology in Figure 2c, the annealing at 150 °C hardly changed the morphology and reflectivity. Although annealing at 200 °C leads to visible coarsening, the arrays remain separate and the reflectivity remains unchanged. As for bare AgNRs, they melted completely at 200 °C and experienced a substantial SERS enhancement degradation of ~90% once the temperature went beyond 100 °C [12]. Therefore, they were not feasible for high-temperature sensing.

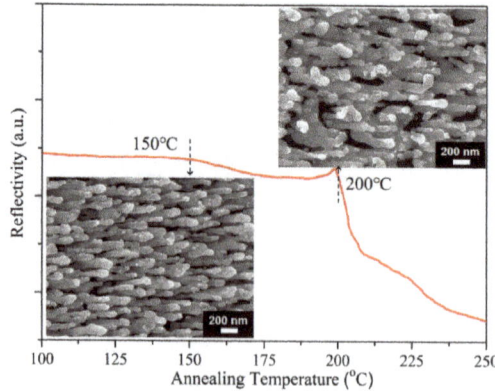

Figure 6. Reflectivity of the AgNRs-Al$_2$O$_3$-Ag arrays as a function of annealing temperature; the insets are SEM images of arrays after annealing at 150 °C (lower left) and 200 °C (upper right), respectively.

Finally, as an application, we investigate the reorientation process of 4-tBBM molecules on this unique AgNRs-Al$_2$O$_3$-Ag nanostructure. We first collected the SERS signals of 4-tBBM molecules on the substrate surface at room temperature, and then heated the analyte-adsorbed substrate at 150 °C for 5 min, followed by measuring the SERS spectra again. Figure 7a presents the SERS spectra of 4-tBBM adsorbed on AgNRs-Al$_2$O$_3$-Ag before and after annealing. At room temperature, there are prominent SERS peaks at 1108 and 1192 cm^{-1} (the vibration of the phenyl ring), 1227 cm^{-1} (the wagging of methylene groups) and 1608 cm^{-1} (8a mode) [27]. Upon heating to 150 °C, the SERS intensities at 1108 and 1192 cm^{-1} increased and the signal at 1227 cm^{-1} remained steady. Of particular interest is that the peak at 1608 cm^{-1} shifted to 1599 cm^{-1}. These intensity and peak position variations imply that the orientation changes of 4-tBBM molecules on the substrate surface during heating. To be specific, the intensity increase at 1192 and 1108 cm^{-1} of in-plane modes of 4-tBBM suggests that the angle between the phenyl ring and the surface normal of the substrate decreases upon annealing [27], a schematic diagram of which is shown in Figure 7b. The peak shift from 1608 to 1599 cm^{-1} also verifies such a transition. Therefore, thermal energy accelerates the molecular vibration and drives 4-tBBM to rearrange to a more stable orientation. AgNRs-Al$_2$O$_3$-Ag thus holds great potential to determine the orientation and conformation of interfacial molecules on SERS substrates.

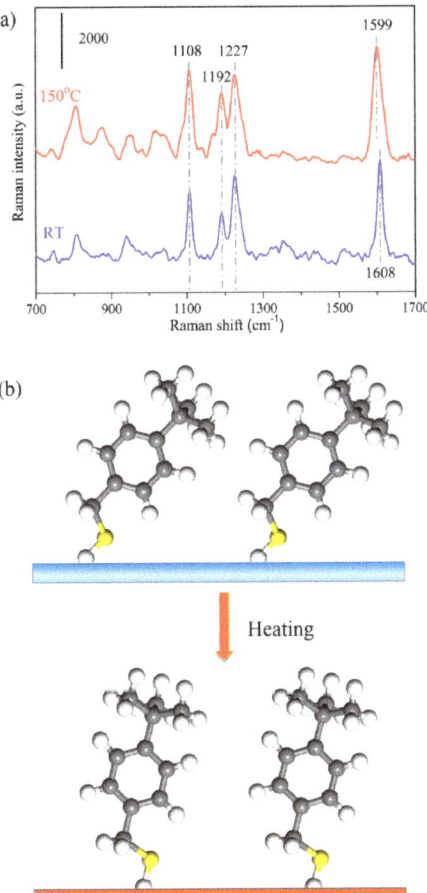

Figure 7. (a) SERS spectra of 1 × 10^{-6} M 4-tBBM molecules adsorbed on AgNRs-Al$_2$O$_3$-Ag arrays with 150 nm additional Ag deposition, at room temperature (RT) and upon heating at 150 °C for 5 min. (b) Schematic of 4-tBBM molecule reorientation process upon heating.

4. Conclusions

In conclusion, through the design of nanoscale heterojunctions, we have developed AgNRs-based SERS substrates that (1) have about 400% times the sensitivity of pure AgNRs and (2) are thermally stable up to 200 °C. The AgNRs-Al$_2$O$_3$-Ag heterojunctions are prepared facilely using physical vapor deposition under the glancing angle incidence. The optimal thickness of additional Ag deposition on top of AgNRs-Al$_2$O$_3$ is 150 nm. To demonstrate the impacts of the design, we have shown that the AgNRs-Al$_2$O$_3$-Ag arrays have an ultrahigh SERS enhancement factor on the order of 10^8. Additionally, the heterojunction arrays allow the characterization of reorientation processes of interfacial 4-tBBM molecules on the substrate upon heating. The AgNRs-Al$_2$O$_3$-Ag arrays, with both superior SERS sensitivity and thermal stability, hold great potential for real-world SERS applications.

Supplementary Materials: The following are available online at http://www.mdpi.com/2079-4991/9/6/830/s1, Figure S1: SEM images of (a) uncapped AgNRs and (b) AgNRs-10Al$_2$O$_3$ substrates; HRTEM images of AgNRs capped with Al$_2$O$_3$ layers of (c) 8, (d) 10 and (e) 20 nm, respectively. Figure S2: (a) SERS spectra of 1×10^{-6} M MB adsorbed on bare AgNRs and on AgNRs-Al$_2$O$_3$ substrates with different capping thickness; (b) The normalized intensities of 1622 cm-1 Raman peak of 1×10^{-6} M MB molecules as a function of the Al$_2$O$_3$ thickness of different AgNRs-Al$_2$O$_3$ substrates. Figure S3: SEM images of AgNRs and different AgNRs-Al$_2$O$_3$ substrates after annealing at 150 °C and 200 °C for 15 min. Figure S4: The reflectivity changes of different AgNRs-Al$_2$O$_3$ substrates upon annealing from 100 °C to 250 °C. Figure S5: SEM images of (a) AgNRs-150Ag and (b) 10Al$_2$O$_3$-150Ag substrates; (c) SERS spectra of 1×10^{-6} M MB molecules adsorbed on AgNRs-10Al$_2$O$_3$-150Ag, AgNRs-150Ag and 10Al$_2$O$_3$-150Ag substrates, separately.

Author Contributions: Conceptualization, H.H. and L.M.; methodology, Z.Z.; software, J.W.; validation, Y.F. and X.L.; formal analysis, L.M.; investigation, Z.Z. and Y.F.; resources, L.M.; data curation, J.W. and X.L.; writing—original draft preparation, L.M.; writing—review and editing, H.H.; visualization, L.M.; supervision, Z.Z. and H.H.; project administration, Z.Z. and H.H.; funding acquisition, Z.Z. and H.H.

Funding: This research was funded by the National Natural Science Foundation of China (51531006); the China Postdoctoral Science Foundation (2018M641189); the Fundamental Research Funds for the Central Universities (FRF-TP-18-090A1). Author Hanchen Huang acknowledges the sponsorship of US Department of Energy Office of Basic Energy Science (DE-SC0014035).

Conflicts of Interest: The authors declare no conflict of interest.

References

1. Fateixa, S.; Nogueira, H.; Trindade, T. Hybrid nanostructures for SERS: Materials development and chemical detection. *Phys. Chem. Chem. Phys.* **2015**, *17*, 21046–21071. [CrossRef] [PubMed]
2. Ma, Y.; Huang, Z.; Li, S.; Zhao, C. Surface-enhanced Raman spectroscopy on self-assembled Au nanoparticles arrays for pesticides residues multiplex detection under complex environment. *Nanomaterials* **2019**, *9*, 426. [CrossRef] [PubMed]
3. Sitjar, J.; Liao, L.; Lee, H.; Liu, B.; Fu, W. SERS-active substrate with collective amplification design for trace analysis of pesticides. *Nanomaterials* **2019**, *9*, 664. [CrossRef] [PubMed]
4. Kalachyova, Y.; Erzina, M.; Postnikov, P.; Svorcik, V.; Lyutakov, O. Flexible SERS substrate for portable Raman analysis of biosamples. *Appl. Surf. Sci.* **2018**, *458*, 95–99. [CrossRef]
5. Chen, H.; Lin, M.; Wang, C.; Chang, Y.; Gwo, S. Large-scale hot spot engineering for quantitative SERS at the single-molecule scale. *J. Am. Chem. Soc.* **2015**, *137*, 13698–13705. [CrossRef] [PubMed]
6. Liu, K.; Bai, Y.; Zhang, L.; Yang, Z.; Fan, Q.; Zheng, H.; Yin, Y.; Gao, C. Porous Au-Ag nanospheres with high-density and highly accessible hotspots for SERS analysis. *Nano Lett.* **2016**, *16*, 3675–3681. [CrossRef] [PubMed]
7. Yang, C.; Zhang, C.; Huo, Y.; Jiang, S.; Qiu, H.; Xu, Y.; Li, X.; Man, B. Shell-isolated graphene@Cu nanoparticles on graphene@Cu substrates for the application in SERS. *Carbon* **2016**, *98*, 526–533. [CrossRef]
8. Bottani, C.; Bassi, A.; Tanner, B.; Stella, A.; Tognini, P.; Cheyssac, P.; Kofman, R. Melting in metallic Sn nanoparticles studied by surface Brillouin scattering and synchrotron-x-ray diffraction. *Phys. Rev. B* **1999**, *59*, 15601–15604. [CrossRef]
9. Pan, C.; Zhang, Z.; Su, X.; Zhao, Y.; Liu, J. Characterization of Fe nanorods grown directly from submicron-sized iron grains by thermal evaporation. *Phys. Rev. B* **2004**, *70*, 233404. [CrossRef]

10. Bachenheimer, L.; Elliott, P.; Stagon, S.; Huang, H. Enhanced thermal stability of Ag nanorods through capping. *Appl. Phys. Lett.* **2014**, *105*, 213104. [CrossRef]
11. Mai, F.; Yang, K.; Liu, Y.; Hsu, T. Improved stabilities on surface-enhanced Raman scattering-active Ag/Al$_2$O$_3$ films on substrates. *Analyst* **2012**, *137*, 5906–5912. [CrossRef] [PubMed]
12. Ma, L.; Zhang, Z.; Huang, H. Design of Ag nanorods for sensitivity and thermal stability of surface-enhanced Raman scattering. *Nanotechnology* **2017**, *28*, 405602. [CrossRef] [PubMed]
13. John, J.; Mahurin, S.; Dai, S.; Sepaniak, M. Use of atomic layer deposition to improve the stability of silver substrates for in situ, high-temperature SERS measurements. *J. Raman. Spectrosc.* **2009**, *41*, 4–11. [CrossRef]
14. Mahurin, S.; John, J.; Sepaniak, M.; Dai, S. A reusable surface-enhanced Raman scattering (SERS) substrate prepared by atomic layer deposition of alumina on a multi-layer gold and silver film. *Appl. Spectrosc.* **2011**, *65*, 417–422. [CrossRef] [PubMed]
15. Ma, L.; Huang, Y.; Hou, M.; Xie, Z.; Zhang, Z. Ag nanorods coated with ultrathin TiO$_2$ shells as stable and recyclable SERS substrates. *Sci. Rep.* **2015**, *5*, 15442. [CrossRef]
16. Ma, L.; Wu, H.; Huang, Y.; Zou, S.; Li, J.; Zhang, Z. High-performance real-time SERS detection with recyclable Ag nanorods@HfO$_2$ substrates. *ACS Appl. Mater. Interfaces* **2016**, *8*, 27162–27168. [CrossRef] [PubMed]
17. Masango, S.; Hackler, R.; Henry, A.; McAnally, M.; Schatz, G.; Stair, P.; Van Duyne, R. Probing the chemistry of alumina atomic layer deposition using operando surface-enhanced Raman spectroscopy. *J. Phys. Chem. C* **2016**, *120*, 3822–3833. [CrossRef]
18. Zhang, J.; Zhang, D.; Shen, D. Orientation study of atactic poly(methyl methacrylate) thin film by SERS and RAIR spectra. *Macromolecules* **2002**, *35*, 5140–5144. [CrossRef]
19. Formo, E.; Wu, Z.; Mahurin, S.; Dai, S. In situ high temperature surface enhanced Raman spectroscopy for the study of interface phenomena: Probing a solid acid on alumina. *J Phys. Chem. C* **2011**, *115*, 9068–9073. [CrossRef]
20. Willets, K.; Van Duyne, R. Localized surface plasmon resonance spectroscopy and sensing. *Annu. Rev. Phys. Chem.* **2007**, *58*, 267–297. [CrossRef]
21. Lai, Y.; Chen, S.; Hayashi, M. Mesostructured arrays of nanometer-spaced gold nanoparticles for ultrahigh number density of SERS hot spots. *Adv. Funct. Mater.* **2014**, *24*, 2544–2552. [CrossRef]
22. Shiohara, A.; Wang, Y.; Liz-Marzán, L. Recent approaches toward creation of hot spots for SERS detection. *J. Photochem. Photobiol. C* **2014**, *21*, 2–25. [CrossRef]
23. Han, C.; Yao, Y.; Wang, W.; Qu, L.; Bradley, L.; Sun, S.; Zhao, Y. Rapid and sensitive detection of sodium saccharin in soft drinks by silver nanorod array SERS substrates. *Sens. Actuators B Chem.* **2017**, *251*, 272–279. [CrossRef]
24. Liu, Y.; Chu, H.; Zhao, Y. Silver nanorod array substrates fabricated by oblique angle deposition: morphological, optical, and SERS characterizations. *J. Phys. Chem. C* **2010**, *114*, 8176–8183. [CrossRef]
25. Driskell, J.; Shanmukh, S.; Liu, Y.; Chaney, S.; Tang, X.; Zhao, Y.; Dluhy, R. The use of aligned silver nanorod arrays prepared by oblique angle deposition as surface enhanced Raman scattering substrates. *J. Phys. Chem. C* **2008**, *112*, 895–901. [CrossRef]
26. Xiao, G.; Man, S. Surface-enhanced Raman scattering of methylene blue adsorbed on cap-shaped silver nanoparticles. *Chem. Phys. Lett.* **2007**, *447*, 305–309. [CrossRef]
27. Tong, L.; Zhu, T.; Liu, Z. Laser irradiation induced spectral evolution of the surface-enhanced Raman scattering (SERS) of 4-tert-butylbenzylmercaptan on gold nanoparticles assembly. *Sci. China Ser. B* **2007**, *50*, 520–525. [CrossRef]

© 2019 by the authors. Licensee MDPI, Basel, Switzerland. This article is an open access article distributed under the terms and conditions of the Creative Commons Attribution (CC BY) license (http://creativecommons.org/licenses/by/4.0/).

Review

In Situ and Real-Time Nanoscale Monitoring of Ultra-Thin Metal Film Growth Using Optical and Electrical Diagnostic Tools

Jonathan Colin [1], Andreas Jamnig [1,2], Clarisse Furgeaud [1], Anny Michel [1], Nikolaos Pliatsikas [2], Kostas Sarakinos [2,*] and Gregory Abadias [1,*]

1. Institut Pprime, UPR 3346, CNRS-Université de Poitiers-ENSMA, 11 Boulevard Marie et Pierre Curie, TSA 41123, CEDEX 9, 86073 Poitiers, France; JJCOLIN@protonmail.com (J.C.); andreas.jamnig@liu.se (A.J.); cfurgeaud@posta.unizar.es (C.F.); anny.s.michel@univ-poitiers.fr (A.M.)
2. Nanoscale Engineering Division, Department of Physics, Chemistry and Biology, Linköping University, SE 581 83 Linköping, Sweden; nikolaos.pliatsikas@liu.se
* Correspondence: kostas.sarakinos@liu.se (K.S.); gregory.abadias@univ-poitiers.fr (G.A.)

Received: 3 October 2020; Accepted: 3 November 2020; Published: 9 November 2020

Abstract: Continued downscaling of functional layers for key enabling devices has prompted the development of characterization tools to probe and dynamically control thin film formation stages and ensure the desired film morphology and functionalities in terms of, e.g., layer surface smoothness or electrical properties. In this work, we review the combined use of in situ and real-time optical (wafer curvature, spectroscopic ellipsometry) and electrical probes for gaining insights into the early growth stages of magnetron-sputter-deposited films. Data are reported for a large variety of metals characterized by different atomic mobilities and interface reactivities. For fcc noble-metal films (Ag, Cu, Pd) exhibiting a pronounced three-dimensional growth on weakly-interacting substrates (SiO_2, amorphous carbon (a-C)), wafer curvature, spectroscopic ellipsometry, and resistivity techniques are shown to be complementary in studying the morphological evolution of discontinuous layers, and determining the percolation threshold and the onset of continuous film formation. The influence of growth kinetics (in terms of intrinsic atomic mobility, substrate temperature, deposition rate, deposition flux temporal profile) and the effect of deposited energy (through changes in working pressure or bias voltage) on the various morphological transition thicknesses is critically examined. For bcc transition metals, like Fe and Mo deposited on a-Si, in situ and real-time growth monitoring data exhibit transient features at a critical layer thickness of ~2 nm, which is a fingerprint of an interface-mediated crystalline-to-amorphous phase transition, while such behavior is not observed for Ta films that crystallize into their metastable tetragonal β-Ta allotropic phase. The potential of optical and electrical diagnostic tools is also explored to reveal complex interfacial reactions and their effect on growth of Pd films on a Si or a-Ge interlayers. For all case studies presented in the article, in situ data are complemented with and benchmarked against ex situ structural and morphological analyses.

Keywords: real-time monitoring; polycrystalline film growth; growth dynamics; interface reactivity; adatom mobility; wafer curvature; electrical resistance; spectroscopic ellipsometry

1. Introduction

Metal films with thicknesses of ~10 nm and below are ubiquitous in many modern life technologies, including microelectronics, displays, sensors, and energy storage/saving/conversion devices [1–8]. Such ultra-thin layers may form a continuous structure and fully wet underlying substrates or self-assemble into discrete nanoscale particles forming a discontinuous morphology. The latter film morphology (i.e., supported nanoparticles) is relevant for the field of heterogeneous catalysis and plasmonics, whereby nanoparticles with high surface-to-volume ratios and unique optical response emanating from localized surface plasmon resonance (LSPR) are leveraged in a broad range of applications [1,9–15]. Concurrently, fabrication of continuous and ultrasmooth metallic layers at thicknesses below 10 nm is desirable for opto-electronic applications which rely on, e.g., conductive transparent electrodes [6,16–18].

A significant fraction of thin films is today synthesized via vapor condensation, a robust and versatile method routinely employed in industry and research laboratories. Material from a solid or liquid source is vaporized using physical and/or chemical means (e.g., by heating or momentum transfer); vapor is transported through the gas phase and condenses on a substrate where it forms a film. During the condensation process, the flux of atoms (and molecules) from the vapor to the solid substrate surface is typically multiple orders of magnitude larger than the flux of material returning from the substrate surface to the vapor phase. This flux difference (also known as supersaturation at the vapor/solid interface) leads to excess of atoms on the substrate, so that film-forming species do not have sufficient time to self-assemble into minimum-energy configurations predicted by thermodynamics. It is then said that film formation proceeds far from thermodynamic equilibrium and the resulting film morphology and microstructure are determined by the occurrence rates (i.e., kinetics) of atomic-scale structure-forming mechanisms [19–25]. Further aspects, crucial for film growth, are chemical reactions (i.e., compound formation) and intermixing at the film/substrate interface, which depend not only on kinetics but are also governed by the thermodynamics (i.e., miscibility vs. immiscibility) of the materials involved [26–29].

Film physical attributes are closely linked with mesoscale morphological and nanoscale structural features, including grain/island size and shape, crystal structure and orientation, and surface roughness. Such features are difficult to predict a priori because they are determined by a complex interplay among a multitude of deposition process parameters, as well as by film/substrate interactions. Hence, the use of robust and non-destructive characterization tools that can provide information at the nanoscale is required for establishing the correlation among atomic-scale mechanisms and resulting film morphology. Implementing such techniques during synthesis (i.e., in situ and in real-time) is particularly advantageous, since it allows one to selectively study dynamic growth processes and decouple them from post-growth microstructural changes.

A wide palette of techniques for in situ thin-film growth monitoring is nowadays available and can be grouped into different categories, based on the measured physical quantities and operation principle (real-space imaging, diffraction, spectroscopy). Scanning probe microscopy (SPM) techniques provide a direct observation of atoms and clusters, as well as of their mobility, through real-space imaging of the film surface electronic density with sub-Ångström vertical resolution. As such, valuable information on atomic-scale mechanisms and their rates (e.g., diffusion barriers) can be obtained. However, SPM techniques are inherently restricted to the characterization of the island nucleation and growth stages at sub-monolayer metal coverage [4,30], require an ultra-high vacuum environment, and data acquisition rate is seldom compatible with a real-time growth monitoring [31,32].

Methods relying on low-energy electron microscopy (LEEM) provide access to mesoscopic lateral length scales (2–150 µm), with video-rate imaging amenable to studying dynamic processes on surfaces, but their lowest lateral resolution is of the order of 10 nm [33–35]. Crystal structure and film growth mode can be studied using reciprocal-space electron-based diffraction techniques, such as low-energy electron diffraction (LEED) or reflection high-energy electron diffraction (RHEED). LEEM, RHEED and LEED are surface-sensitive probes that they are ideally suited to study growth up to few monolayers (ML). They require ultra-high vacuum conditions and, hence, they are usually implemented to investigate thin film growth by thermal evaporation, although experimental setups based on differential pumping stages can be designed to be compatible with sputter-deposition. The aforementioned SPM and electron scattering techniques are often not directly integrated into deposition chambers; hence analysis is performed in a "stop-and-growth" fashion, in which a certain amount of metal deposit is iteratively probed at key film formation stages in separate analysis chambers.

Non-invasive, surface sensitive techniques based on X-rays can be advantageously employed to study the structural, morphological and chemical evolution during thin film growth [36–43]. X-ray reflectivity (XRR), grazing incidence small-angle X-ray scattering (GISAXS), X-ray diffraction (XRD), X-ray fluorescence (XRF), and X-ray absorption spectroscopy (XAS) can be used remotely to probe the sample surface, with the only requirement being that the deposition chamber must be equipped with X-ray transparent windows (such as beryllium or Kapton). These methods can be used separately or coupled to each another, but most in situ experiments during film deposition require synchrotron-based X-rays. The high-brilliance of third-generation synchrotron sources, along with modern fast two-dimensional (2D) X-ray detectors, facilitate the monitoring of the kinetics of thin film growth in real-time with fast-acquisition (milliseconds) and sub-monolayer precision. Besides, the ability to tune the photon energy to a specific experiment and material system is an additional asset.

Another category of in situ and real-time diagnostics is based on measuring the change of electrical and optical properties of the deposited layer as a function of time. Evolution of electrical properties (e.g., film resistivity) can be measured using four-point probe techniques [44–47], while typical optical diagnostics include reflectance spectroscopy [48–51] and spectroscopic ellipsometry [52–60]. These techniques can characterize all relevant film-growth stages up to the formation of a continuous layer and beyond, while they provide morphological information over mesoscopic length scales. They are also characterized by conceptual and practical simplicity, they are readily available in and compatible with typical thin-film synthesis apparatuses, and data interpretation is in most cases straightforward.

In the present review article, we demonstrate the strength of combining laboratory-scale electrical and optical in situ and real-time diagnostic tools for shedding light onto morphological evolution, structure formation, and growth dynamics in a wide gamut of film/substrate systems, whereby films are grown by physical vapor deposition techniques. In a first group of film/substrate systems, we study Ag, Cu, and Pd growth (all exhibiting fcc crystal structure) on a number of substrates, including Si covered with its native oxide layer, Si covered with a thermally grown SiO_2 layer, and amorphous carbon (a-C). These film/substrate combinations exhibit minimum chemical interactions and reactivity, which allows us to selectively study the effect of atomic-scale kinetics on film growth. An additional effect of the weak film/substate interaction in the latter systems is that the deposited layers grow in a pronounced three-dimensional (3D) fashion, which offers an ideal test bed for identifying subtle changes of film morphology as a function of deposition conditions and material characteristics using optical and electrical probes [1,61,62]. As such, kinetics is studied both in terms of *intrinsic* atomic mobility of the thin-film materials—as approximated by their melting point T_m, which yields homologous temperatures $T_h = T/T_m$ of 0.24 (Ag), 0.22 (Cu), 0.16 (Pd and Fe), 0.1 (Mo) and 0.09 (Ta), at $T = 300$ K, where T is the substrate temperature [21,63–65]—and *extrinsic* deposition parameters, including deposition temperature and rate. In a second group, the importance of interface reactivity on film structure formation is addressed by discussing the growth of bcc transition metals (Fe, Mo and Ta) on amorphous Si (a-Si). Interfacial reactions are further examined by monitoring the growth of Pd films on a-Si and a-Ge layers. The selected metal/Si systems span a wide range of chemical reactivities. Although interface reaction and silicide formation during thermal annealing is well documented in the literature [26,66], studies on the nucleation processes during metal deposition on a-Si are scarce.

The content of the article is predominantly based on results generated by us over the past years and focused on metal films synthesized by magnetron sputtering using Ar plasma discharges. Our results are critically complemented by literature data, in order to expand the scope and relevance of our conclusions.

The article is organized as follows: Section 2 explains the overall strategy for thin-film synthesis and provides a brief description of the techniques used for in situ and real-time growth monitoring; Section 3 demonstrates the use of in situ an real-time techniques for studying morphological evolution and growth dynamics of metals on weakly-interacting substrates; Section 4 addresses the effect of interfacial reactivity on film morphological evolution as established by in situ and real-time methodologies; Section 5 summarizes the article and presents an outlook for future developments in the field.

2. Film Synthesis and Real-Time Growth Monitoring

2.1. Film Synthesis and In Situ/Real-Time Monitoring Strategy

Thin-film growth was performed by means of magnetron sputtering in three vacuum chambers located at the University of Poitiers (France) and Linköping University (Sweden), all equipped with a load-lock sample transfer system and multiple cathodes arranged in a confocal configuration. Moreover, the vacuum chambers are specifically designed to host techniques for real-time monitoring of the deposition process. One of the film deposition setups (at the University of Poitiers) achieves high-vacuum conditions (base pressure $\leq 8 \times 10^{-6}$ Pa using a cryogenic pump); it enables measurements of stress evolution during film growth using the wafer curvature method (details are given in Section 2.2), while it also allows for monitoring the change in the film electrical resistance via the four-point probe technique (details are given in Section 2.3) using a custom-built sample holder stage. We note that, due to geometrical constraints and specificity of sample dimensions, the two diagnostics cannot be used simultaneously. The other two deposition chambers (at Linköping University) achieve ultra-high-vacuum conditions (base pressures ~10^{-7}–10^{-8} Pa using turbomolecular pumps) and feature transparent viewports for mounting a spectroscopic ellipsometer to monitor the change in optical response of the growing layer (see Section 2.4 for details). All vacuum systems are equipped with a resistive substrate heater, such that deposition temperature can be varied and temperature effects on film-forming processes can be investigated. Deposition flux is controlled by changing the electrical power applied to the magnetrons. The generic layout of the deposition apparatuses, along with the in situ diagnostic tools, is schematically depicted in Figure 1.

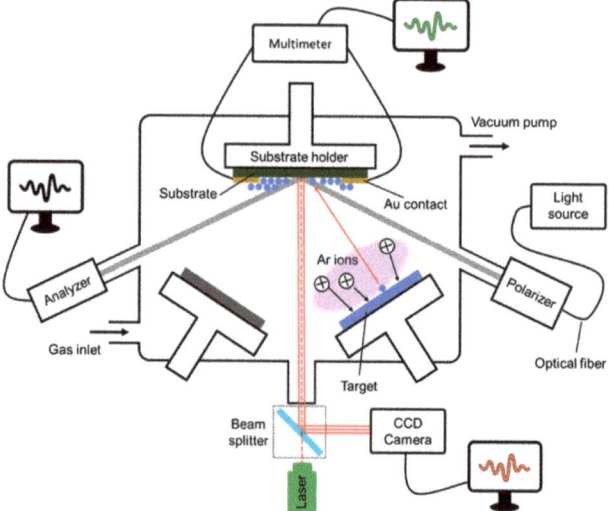

Figure 1. Generic schematic illustration of the sputter-deposition chamber used for collecting data reported in this work. The chamber is equipped with several in situ diagnostics which allow real-time growth monitoring: the wafer curvature set-up is attached at the bottom flange of the chamber and consists of a multiple-beam laser illuminating the substrate at near-normal incidence; the set-up for spectroscopic ellipsometry, operating at an incidence angle of ~70°, consists of a light source, polarizer, and analyzer. The sample holder stage can be fitted with a custom-built four-point probe apparatus to measure the change in electrical resistance during deposition.

2.2. Wafer Curvature Method

The wafer curvature method is based upon measuring the variation of the substrate curvature $\Delta\kappa$ induced by the existence of stress in the film that is attached to it. There are different ways to detect the change in curvature, but the most sensitive and easy to implement in situ during deposition is the method relying on the optical measurement of $\Delta\kappa$ using laser deflectometry [67,68]. In this work, we report data that were recorded with a multiple-beam optical stress sensor (MOSS) set-up, designed by k Space Associates (Dexter, MI, USA) [69,70]. The main advantage of illuminating the sample with multiple beams simultaneously is to alleviate the sensitivity to ambient vibrations during data acquisition: when using a beam array, $\Delta\kappa$ is calculated by measuring the relative spacings $\Delta d = d - d_0$ between adjacent spots, instead of recording the absolute position of one reflected beam. A 3 × 3 array of parallel beams, with initial spacing d_0, is created using two etalons (beam splitters), and the beams reflected off the substrate are detected on a CCD camera located at a distance L from the substrate. A dedicated software allows for accurate measurement of the variation in spot spacing $d(t)$ as a function of time t with typical acquisition rate of 10 Hz. The change in curvature $\Delta\kappa$ is then obtained from the expression

$$\Delta\kappa(t) = \frac{\cos\alpha}{2L}\frac{\Delta d}{d_0} = \frac{\cos\alpha}{2L}\left(\frac{d(t) - d_0}{d_0}\right) \quad (1)$$

where α is the incidence angle of the laser beam with respect to the substrate normal. In the curvature measurement setup used in the present work, the laser illuminates the substrate at near-normal incidence ($\alpha\sim 0°$) and $L\sim 70$ cm. It is noted that the accurate determination of the optical distance L is realized using a set of mirrors with known curvature.

The biaxial stress in the growing film at a distance (i.e., height) z from the film/substrate interface, $\sigma(z)$, is directly obtained from $\Delta\kappa$ using Stoney's equation according to [70]

$$\langle\sigma_f\rangle \times h_f = \int_0^{h_f} \sigma(z)dz = \frac{M_s h_s^2}{6}\Delta\kappa \quad (2)$$

where $\langle\sigma_f\rangle \times h_f$ is the stress-thickness product (also referred to as the force per unit width, expressed in N/m), $\langle\sigma_f\rangle$ is the average stress in the film at thickness h_f, and h_s and M_s are the thickness and biaxial modulus of the substrate, respectively. For the MOSS measurements presented herein, 100 ± 2 µm thick Si (001) substrates (with dimensions of 1 × 1 cm^2) were used. The substrates were mounted loosely on a sample holder, such that free bending during growth is possible.

In the present work, the substrate curvature method is not merely used as a stress evaluation technique but also as a sub-nanometer-scale sensitive tool for real-time monitoring of film/substrate interfacial reactions, island nucleation, island coalescence, and overall film morphological evolution [71,72]. For instance, thin films growing in a 3D fashion exhibit a characteristic compressive-tensile-compressive stress evolution with increasing film thickness, and the position of the tensile peak maximum has been shown to coincide with the thickness h_{cont} at which a continuous layer is formed [73].

2.3. Electrical Resistance

The second in situ diagnostic that is implemented to monitor film growth evolution is a custom-built apparatus with four-point probe (4PP) arrangement for measuring the variation of the film sheet resistance R_s during deposition. The setup consists of a sample-holder stage, compatible with transfer system from the load-lock to the main deposition chamber, and an electrical collector mounted in the main chamber. The collector is equipped with feedthrough connectors to a Keithley sourcemeter that is interfaced to a PC and controlled by a dedicated software. This setup allows measurements on a series of samples without venting the main chamber. The stainless-steel substrate holder is insulated from the substrate using a 6 mm thick Teflon disk, which can be heated from the back side using a resistive heater. Gold contacts are pre-deposited on the Si substrate, and a conical mask is used during metal film growth to protect the contacts from the vapor flux. More details on the in situ resistivity set-up can be found in [47]. Growth is performed on 350 µm thick, highly-resistive (with resistivity in the range 1–5 kΩ·cm) Si wafers to maximize the change in electrical resistance upon metallic film deposition. We report here the evolution of $R_s \times h_f$ vs. deposition time (or film thickness h_f). Note that the quantity $R_s \times h_f$ is proportional to the film resistivity ρ, which is derived by applying a correction factor to account for the specific sample geometry. In this work, since the sample geometry remains unchanged, we will only report the raw data in the form of $R_s \times h_f$ vs. h_f curves, from which two morphological transition thicknesses, i.e., the percolation (h_{perc}) and the continuous formation (h_{cont}) thicknesses are extracted.

2.4. Spectroscopic Ellipsometry

Spectroscopic ellipsometry (SE) is a non-destructive optical technique in which linearly or circularly polarized light is used to irradiate the sample under investigation [74]. Upon interaction (i.e., reflection or transmission) with the sample, the polarization state of light becomes elliptical. By measuring the change in the light polarization state, the optical properties of the sample can be determined. Figure 2a depicts schematically the concept of ellipsometry for the cases of linearly polarized incident light and reflection geometry. To describe the change in polarization, the reflectance is analyzed into the orthogonal s-p system, where s and p denote planes that are parallel and perpendicular to the plane of incidence, respectively. By measuring the reflected intensity (i.e., the intensity of the electric field $\vec{E}$) along the s and p directions, the ellipsometric angles Ψ and Δ (amplitude ratio and phase shift, respectively, of the reflected light relative to the incident light) are determined. In the case of a bulk sample (i.e., a sample in which the incident light is only absorbed from and reflected at the sample/ambient interface), the complex dielectric function $\tilde{\varepsilon}(\omega)$ of the material under investigation can be computed directly from the quantities Ψ and Δ. The latter is not possible when the sample consists of a partially transparent film residing on the substrate, as in that case Ψ and Δ depend in a non-trivial fashion on the optical response of the substrate and the film, as well as on the film thickness. Hence, the use of models is required for determining the optical properties of the thin film, as shown schematically in Figure 2b and explained hereafter.

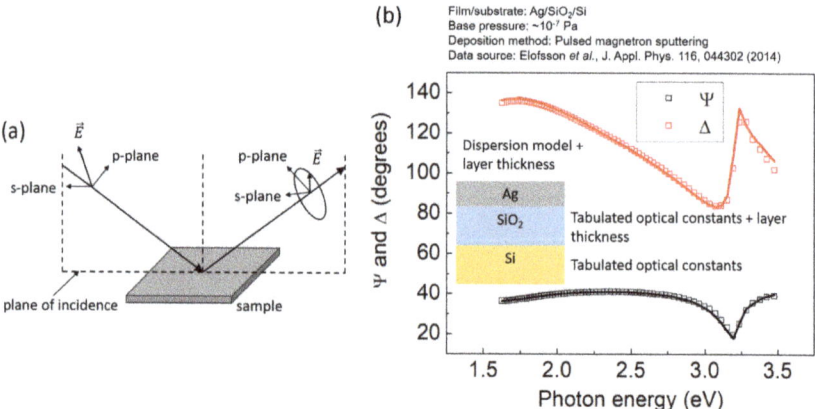

Figure 2. (a) Schematic illustration of the principle of spectroscopic ellipsometry. Linearly polarized light (with electric field vector $\vec{E}$) is reflected at sample surface. Reflection of the incident light causes change for the polarization state to elliptical. p-plane and s-plane indicate the planes that are parallel and perpendicular to the plane of incidence, respectively. By measuring the reflected intensity (i.e., intensity of the electric field $\vec{E}$) along the s and p directions, the ellipsometric angles Ψ and Δ (amplitude

ratio and phase shift, respectively, of the reflected light relative to the incident light) are determined from which the optical properties of the sample under investigation can be extracted. (**b**) Ψ and Δ experimental data (square symbols) measured from an electrically conductive Ag film grown on Si substrate covered by thermally grown ~300 nm SiO$_2$ layer (data are taken from [62]). The ellipsometric data depend in a complex manner on the optical properties of Si, SiO$_2$, and Ag, as well as on the SiO$_2$ and Ag layer thicknesses. Data are fitted using three-phase model (model data are represented by solid lines) which is schematically depicted in the inset with additional details provided in the text.

In situ and real-time SE is used to monitor the evolution of the angles Ψ and Δ over multiple wavelengths during film growth and, by using appropriate models, determine the changes of the optical properties of the deposited layer. These changes are then correlated with the overall film morphological evolution, as explained in detail in Section 3.

The model system that is used in the present article for demonstrating the ability of SE to study film growth is Ag/SiO$_2$/Si. For such films, ellipsometric angles are acquired every ~2 s at 67 incident-light photon energies in the range 1.6–3.2 eV, at an angle of incidence of ~70° from the substrate normal (see representative curves in Figure 2b from [62]). The acquired data are fitted to a three-phase model consisting of substrate, film, and ambient (see Figure 2b). The substrate is modeled as a 625 μm-thick Si slab with a SiO$_2$ overlayer, the thickness of which (in the range ~300 to 500 nm) is confirmed by measuring the optical response of the substrate prior to deposition. Reference data for the substrate layers are taken from Herzinger et al. [75]. The optical response of the film is described by the following dispersion models [76] depending on the film growth stage.

Discontinuous layer: During initial growth stages, the Ag films on SiO$_2$ surface primarily self-assemble in discrete islands that support LSPR. Being a resonant effect, LSPR can be described by adapting the Lorentz oscillator model [55,77] to express the complex dielectric function of the layer $\widetilde{\varepsilon}(\omega)$ as

$$\widetilde{\varepsilon}(\omega) = \frac{f\omega_0^2}{\omega_0^2 - \omega^2 - i\Gamma\omega} \tag{3}$$

In Equation (3), f and ω_0 are the oscillator strength and resonance frequency, respectively, and Γ represents the damping rate of the plasmon resonance. The position of LSPR ω_0 is used to gauge changes of film morphology, including changes in substrate area coverage and island size (see Section 3.4).

Electrically conducting layer: The optical response of electrically conductive Ag films is described by the Drude free electron theory, which is extensively used for ideal metals [76]. In this case $\widetilde{\varepsilon}(\omega)$ is given by the expression,

$$\widetilde{\varepsilon}(\omega) = \epsilon_\infty - \frac{\omega_p^2}{\omega^2 + i\Gamma_D\omega} \tag{4}$$

In Equation (4), ϵ_∞ is a constant that accounts for the effect of interband transitions occurring at frequencies higher than the ones considered here, Γ_D is the free-electron damping constant, and $\omega_p = \sqrt{ne^2/\varepsilon_0 m_e}$ is the free-electron plasma frequency, where n is the free-electron density, ε_0 is the permittivity of free space, and m_e is the free-electron effective mass. From Equation (4), the room-temperature film resistivity is calculated as

$$\rho = \frac{\Gamma_D}{\epsilon_0 \omega_p^2} \tag{5}$$

The evolution of resistivity as function of film thickness (the latter is also determined from SE) provides information with regards to continuous layer formation, the degree of 3D clustering and the dynamics of film growth.

3. Growth of Metal Films on Weakly-Interacting Substrates

3.1. Film Growth Stages and Morphological Transitions

The present section provides a brief description of formation stages and morphological evolution during polycrystalline film growth, with emphasis on weakly-interacting film/substrate systems. Growth starts with adsorption of vapor atoms (referred to as adatoms) on the substrate surface and formation of spatially separated single-crystalline islands via agglomeration of adatoms (nucleation), which grow in size (island growth) and impinge on each other forming new larger islands (coalescence). The process of coalescence also leads to a reduction of the island number density on the substrate surface and continues until the boundaries between single-crystalline islands (i.e., grains) become immobile, such that coalescence stops and a network of interconnected polycrystalline islands forms. Subsequent deposition fills the inter-island space with material (hole filling) and, once this process is completed, a continuous film is formed. The afore-mentioned stages can be visualized in Figure 3 which displays the sequence of transmission electron micrographs taken at various nominal thickness during sputter-deposition of Ag and Cu films on SiO_2 and a-C substrates, respectively [78,79]. We note here that the nominal thickness h_f corresponds to the amount of vapor deposited on the substrate surface at any given time t (irrespective of whether the film is discontinuous or continuous), and it is calculated as $h_f = F \times t$ with F being the deposition rate as determined by the thickness of a continuous layer (e.g., from XRR).

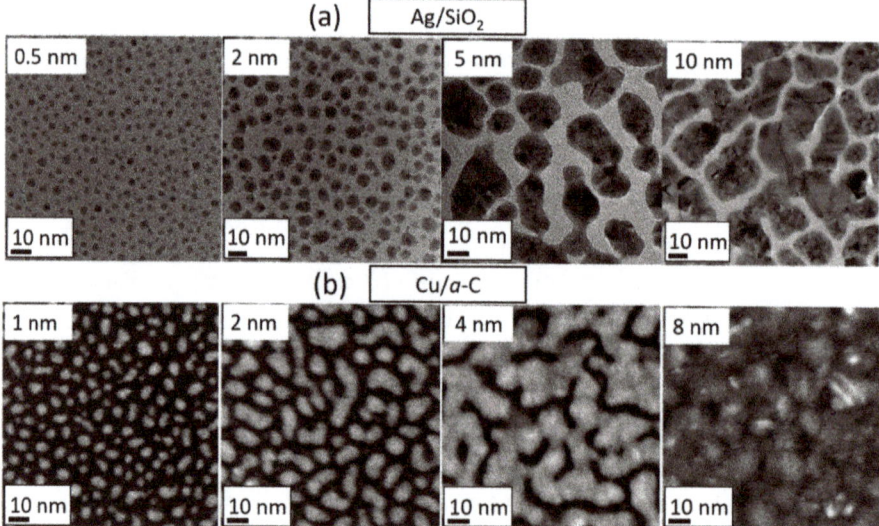

Figure 3. Plan-view TEM micrographs showing the morphological evolution and various formation stages of (**a**) Ag and (**b**) Cu thin films with different thickness deposited on SiO_2 (Ag) and a-C (Cu) by magnetron sputtering at room temperature. Island nucleation and growth at 0.5 nm in (**a**); complete coalescence at 2 nm in (**a**); incomplete coalescence and formation of elongated islands at 5 and 10 nm in (**a**) and 1 and 2 nm in (**b**); hole-filling at 4 nm in (**b**); continuous-layer formation at 8 nm in (**b**). Micrographs in (**a**) correspond to bright-field images (Reprinted with permission from [79]. Copyright ACS 2020), while the images in (**b**) were taken using scanning transmission electron microscopy in high-angle annular dark field (STEM-HAADF) mode (Reprinted with permission from [78]).

Throughout the various film formation stages, competing atomic-scale processes are operative, giving rise to characteristic morphological transitions, which provide information on the degree of

3D clustering (which is inherent in weakly-interacting film/substrate systems) and the overall growth dynamics. These transitions are explained in the following.

Island density saturation: At finite temperatures, adatoms perform a two-dimensional random walk on the substrate surface with a diffusivity D, a quantity that depends on the potential energy landscape encountered by the adatoms and on the growth temperature. Vapor deposition increases the adatom number density on the substrate surface until adatom-adatom encounters lead to nucleation, i.e., the formation of stable atomic clusters (islands) [22]. Nucleation results in an increase of the island number density N on the substrate, until a saturation value N_{sat} is reached. The magnitude of N_{sat} is governed by the competition among formation of new islands and incorporation of adatoms to existing ones, and it is expressed as

$$N_{sat} \propto \left(\frac{F}{D}\right)^x \qquad (6)$$

where $x = \frac{1}{3}\left(\frac{2}{7}\right)$ for 2D (3D) islands [22,80,81]. Increase of (e.g., caused by increasing the deposition temperature T) leads to larger adatom mean free path on the substrate surface. This favors adatom incorporation into existing islands, at the expense of nucleating new ones, and it results in a decrease of N_{sat} Conversely, increase of F leads to a larger adatom number density on the substrate surface. This increases the probability of adatom–adatom encounters and, hence, promotes island nucleation at the expense of island growth, resulting in a larger N_{sat}.

Elongation transition: Islands grow larger by incorporation of adatoms and/or material from the vapor phase. Beyond N_{sat}, island growth becomes the main process that determines film morphology by increasing the fraction of substrate surface covered by the deposit. This is until two or more islands impinge and coalesce into a larger single-crystalline island, which largely erases morphological features attained during earlier stages of film growth. The time required for the coalescing islands to re-establish equilibrium shape (i.e., time for coalescence completion) increases with increasing island radius (i.e., size) R [82,83], until it becomes longer than the time required for a third island to impinge on a coalescing island pair. This point during growth corresponds to the so-called elongation transition, beyond which the film surface consists predominantly of elongated non-coalesced clusters of islands [84]. Analytical modelling, based on the droplet growth theory [85,86], and kinetic Monte Carlo simulations [87–90] suggest that, for film materials and deposition parameters for which coalescence is the dominant process during early stages of film growth (coalescence-controlled growth regime), the nominal film thickness at the elongation transition h_{elong} scales with F (for the case of 3D growth) as

$$h_{elong} \sim \left(\frac{B}{F}\right)^{\frac{1}{3}} \qquad (7)$$

In Equation (7) B is the so-called coalescence strength, which is a material- and temperature-dependent constant [82,83]. Equation (7) reflects the effect of dynamic competition among island growth and coalescence on film morphological evolution. For a constant coalescence strength B, increase of F yields a larger island growth rate, such that an elongated surface morphology is attained at smaller nominal thicknesses. Conversely, an increase of B, at a constant F, promotes coalescence completion relative to island growth, thereby delaying the occurrence of elongation transition.

For a given film/substrate system, there are deposition conditions in terms of F and T, for which coalescence is not completed throughout all stages of growth (coalescence-free growth regime) [87–90]. In this case, h_{elong} becomes proportional to the island-island separation distance when island density reaches N_{sat} [87–90], i.e., $h_{elong} \sim N_{sat}^{-\frac{1}{2}}$. Using Equation (6) for 3D growth, the following expression is obtained:

$$h_{elong} \sim \left(\frac{D}{F}\right)^{\frac{1}{7}} \qquad (8)$$

Equation (8) represents the way by which the interplay among island nucleation and growth (in case that coalescence completion is inactive) affects the early-stage film morphology. An increase of D, for a given F, favors the growth of existing islands, at the expense of nucleation of new ones, resulting in an increase of the nominal thickness required for the onset of island-island impingement. In the opposite case, larger F, at a constant D, promotes nucleation, pushing elongation to occur at smaller nominal thicknesses.

Percolation transition and continuous-layer formation: The onset of elongation transition leads to a film surface that is predominantly covered by polycrystalline islands. The shapes of grain boundaries in these islands change continuously as result of the competition between boundary and surface energies [23], while grain boundaries can be mobile, depending on the growth temperature and the grain size [21]. These effects cause grain growth, which in combination with the kinetically controlled rate at which adatoms descend from the surface of the film to the grain boundary base (hole filling) leads to a formation of an interconnected island network (percolation transition) and eventually to a continuous film. Intuitively, it should be expected that the nominal thickness at which percolation transition occurs (h_{perc}) and a continuous film is formed (h_{cont}) are affected only by the rates of hole filling vs. out-of-plane film growth. However, the influence of initial growth stages (island nucleation, growth, and coalescence) on the values and scaling behavior of those thicknesses is very pronounced, as explained in Section 3.2.

3.2. Experimental Determination of Morphological Transition Thicknesses

Initial growth stages related to island nucleation are typically studied by scanning tunneling microscopy (STM) [22], which is an ideal tool for investigating morphology in epitaxial film/substrate systems, including metals deposited on oxide surfaces [1,30]. However, due to the inherent complexity of STM techniques, most studies of metal film growth on weakly-interacting substrates focus on post-nucleation morphological transitions (elongation, percolation, continuous-film formation) and their respective nominal thicknesses. The absolute value of the elongation transition thickness h_{elong} for a given set of synthesis conditions reflects the degree of 3D clustering during growth, whereby larger h_{elong} indicates a more pronounced 3D morphology. Concurrently, the scaling behavior of h_{elong} as a function of the deposition rate F describes the relative importance of nucleation vs. coalescence for film morphological evolution [89] (see Equations (7) and (8)). The elongation transition is an intrinsically abstract concept, i.e., h_{elong} is difficult to determine experimentally [91]. Hence, subsequent morphological transition thicknesses, i.e., h_{perc} and h_{cont}, are typically measured, as these thicknesses have been shown to scale linearly with h_{elong} [87,88,91].

The formation of an interconnected network of islands (i.e., percolation transition) leads to the onset of electrical conductivity, when a film is deposited on a substantially insulating substrate. Hence, h_{perc} can be determined by measuring in situ the resistivity change of the deposited layer (using the four-point-probe technique; see Section 2.3), whereby h_{perc} corresponds to the thickness at which the measured resistivity exhibits a sharp drop. An example is shown in Figure 4, which plots a $R_s \times h_f$ vs. h_f curve from an Ag film grown by magnetron sputtering on SiO_2 (red solid line; percolation thickness marked with a red solid arrow) [61]. With increasing film thickness beyond h_{perc}, the resistivity decreases further until the $R_s \times h_f$ vs. h_f curve reaches a steady-state (marked by the intersection of the dashed lines and indicated with a red solid arrow). Multiple studies [61,62,92–94], which have combined in situ growth monitoring and ex situ morphology and structure characterization, have shown that the thickness at which steady-state film resistivity is established corresponds to h_{cont}. An alternative approach for determining film resistivity is indirectly by SE using the Drude model, as explained in Section 2.4. A resistivity ρ vs. h_f curve, determined by SE, for Ag grown by magnetron sputtering on SiO_2 is also plotted in Figure 4 (black hollow squares), and h_{cont} (i.e., onset of steady-state behavior) is marked with a black solid arrow at the intersection between the two dashed solid lines used as a guide to the eye. We note here that accuracy of the Drude model close to the onset of conductivity is limited, hence h_{perc} cannot be determined with precision using SE [62].

Stress evolution is also closely connected with the film formation stages [95]. Figure 4 plots the stress-thickness $\langle \sigma_f \rangle \times h_f$ vs. h_f curve of an Ag layer grown by magnetron sputtering on SiO_2 (blue solid line) [61]. The curve exhibits the typical compressive-tensile-compressive (CTC) stress evolution for films grown at conditions of high atomic mobility on weakly-interacting substrates [67,70,95]. The origin of the different stress stages has been the focus of extensive experimental and theoretical works in the literature. It is widely accepted that the first compressive stage corresponds to the nucleation of isolated islands [95,96], while their coalescence leads to tensile stress formation (attractive forces) due to elastic strain upon impingement of neighboring surfaces (similar to a zipping process, see [97,98]), the driving force being the reduction in surface/interface energy upon formation of grain boundary between coalescing islands pairs. As coalescence progresses, the film continues to develop tensile stress up to an observable maximum, which has been shown to coincide with the formation of a continuous layer [73,99]. Therefore, the stress monitoring during thin film growth using MOSS allows the straightforward determination of h_{cont} from the position of the tensile peak maximum, as indicated by the blue arrow in Figure 4. The underlying mechanisms for the origin of the compressive stress in the continuous film growth regime are still the subject of debate [100–104], and will not be further discussed here, as they fall outside the focus of the present review paper. We also note that the differences in the morphological transition thicknesses established by the curves in Figure 4 reflect differences in growth kinetics, as determined by the deposition conditions (e.g., deposition rate, temperature, base and working pressure). A more detailed discussion on this aspect is provided in Section 3.3.

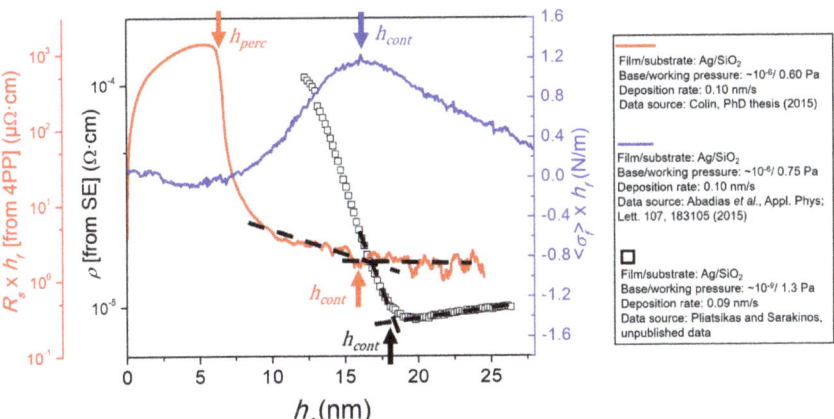

Figure 4. Evolution of resistivity ($\rho \sim R_s \times h_f$) and stress-thickness product $\langle \sigma_f \rangle \times h_f$ vs. nominal film thickness h_f during growth of Ag on SiO$_2$ by magnetron sputtering, as measured from in situ and real-time diagnostic tools: R_s (from four-point-probe measurements; red solid line; data taken from [105], ρ (from spectroscopic ellipsometry (SE); hollow black squares; unpublished data by Pliatsikas and Sarakinos), and $\langle \sigma_f \rangle \times h_f$ (from multiple-beam optical stress sensor (MOSS) measurements; data taken from [73]). Growth conditions (deposition rate and base/working pressure) are indicated in the corresponding legends, while all films have been grown at room temperature. The percolation (h_{perc}) and continuous film formation (h_{cont}) thicknesses are determined by the curves as explained in the text.

3.3. The Effect of Growth Kinetics on Film Morphological Evolution

3.3.1. Influence of Material Intrinsic Mobility

In the present section, we demonstrate the ability of the in situ and real-time techniques described in Section 2, to establish the effect of growth kinetics, as determined by synthesis conditions and intrinsic material properties, on the degree of 3D clustering in weakly-interacting film/substrate systems. We start by examining the evolution of $\langle \sigma_f \rangle \times h_f$ (Figure 5a) and $R_s \times h_f$ (obtained from four-point-probe measurements; Figure 5b) during growth of Ag, Cu, and Pd on SiO$_2$, at otherwise identical deposition conditions (see Table 1 for deposition conditions). The $\langle \sigma_f \rangle \times h_f$ curves exhibit the characteristic CTC stress evolution as a function of h_f, from which the continuous film formation thickness h_{cont} (i.e., tensile peak maximum) is determined for each metal. The $R_s \times h_f$ curves show the abrupt drop at h_{perc}, while the onset of a steady-state signifies h_{cont} (all transition thicknesses are marked with vertical dashed lines in the respective curves in Figure 5). The values of h_{perc} (from $R_s \times h_f$ vs. h_f curves) and h_{cont} (from $\langle \sigma_f \rangle \times h_f$ vs. h_f curves) are listed in Table 1, where it is seen that Ag exhibits the largest values for both quantities (h_{perc} = 5.9 nm and h_{cont} = 12.4 nm), followed by Cu (h_{perc} = 2.6 nm and h_{cont} = 8.2 nm), while Pd has the smallest values with h_{perc} = 1.7 nm and h_{cont} = 5.9 nm. These differences in morphological transition thicknesses indicate that Ag grows in the most pronounced 3D fashion (among the three metals), while Pd exhibits the flattest morphology, as confirmed by ex situ studies of the surface topography of the three metals, see Figure 5c. These findings are also consistent with the TEM observations of Figure 3, where Cu islands form more elongated structures and percolate at lower film thickness compared to Ag. Concurrently, the three metals have distinctly different melting points T_m, so that their homologous temperatures $T_h = T/T_m$ (T is the deposition temperature that is 300 K, common for all metals) are 0.24 (Ag), 0.22 (Cu), and 0.16 (Pd) (also listed in Table 1). Atomic mobility, in a first approximation, scales with T_h [21], i.e., Ag exhibits the largest mobility. This allows Ag to: (i) diffuse longer distances on the substrate and self-assemble into larger and fewer nuclei; (ii) exhibit more pronounced upward diffusion from the base to the top of atomic islands [106,107]; and (iii) diffuse faster on Ag islands so the coalescence completion is promoted [108]. All the aforementioned effects favor 3D growth morphology, and thereby, yield the largest h_{perc} and h_{cont} values. Using the same argument, we can identify the reason for the smallest h_{perc} and h_{cont} values observed for Pd (i.e., less pronounced 3D morphology) to the smaller atomic mobility. Similar findings were reported by Abermann et al. during thermal evaporation of Ag, Au and Cu films [109].

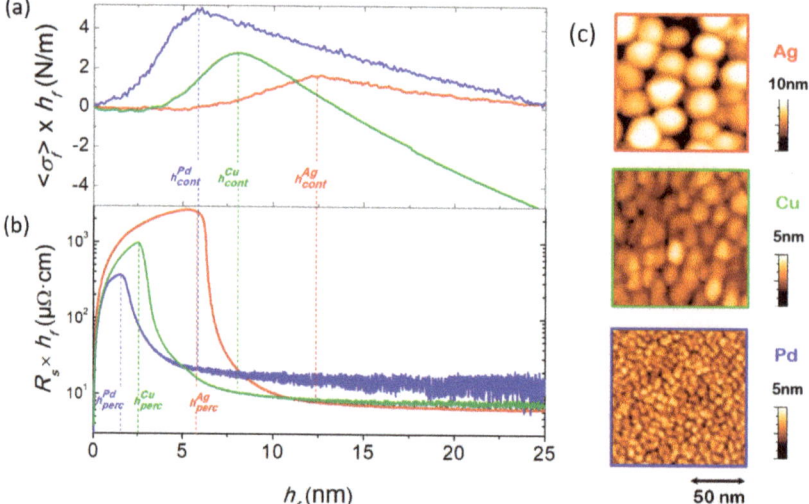

Figure 5. Real-time evolution of (a) $\langle \sigma_f \rangle \times h_f$ and (b) $R_s \times h_f$ vs. h_f for sputter-deposited Ag (red solid line), Cu (green solid line) and Pd (blue solid line) films on SiO_2 at T = 300 K. Morphological transition thicknesses h_{perc} and h_{cont} are indicated on the respective curves with vertical dashed lines of the same color. (c) Atomic force microscopy (tapping mode) images (125 × 125 nm^2) showing the surface morphology of Ag, Cu and Pd films with h_f~3 nm. More information on the growth conditions for each data set is provided in Table 1. Data taken from [71,78].

Table 1. Deposition conditions and characteristic morphological transition thicknesses for Ag, Cu and Pd films deposited by magnetron sputtering at $T = 300$ K on SiO_2. The Ar working pressure is 0.3 Pa.

Element	Crystal Structure	Target Power (W)	Deposition Rate F (nm/s)	T_h	h_{perc} (nm)	h_{cont} (nm)
Ag	fcc	15	0.06	0.24	5.9 ± 0.1	12.4 ± 0.1
Cu	fcc	30	0.06	0.22	2.6 ± 0.1	8.2 ± 0.1
Pd	fcc	30	0.08	0.16	1.7 ± 0.1	5.9 ± 0.1

Atomic mobility is not the sole factor that governs the stress and morphological evolutions. Structure formation (crystalline phase) is another aspect that needs to be investigated. In this example, the three metals crystallize in their fcc structure with (111) preferred orientation, but there are scenarios in which nucleation of a specific crystallographic phase is influenced by interfacial effects. This will be discussed in Section 4.1. Another aspect related to Pd is its reactivity with the Si substrate, which may increase interaction strength, change interface chemistry, and suppress 3D growth. This issue is further examined in Section 4.2.

3.3.2. Influence of Deposition Rate F and Temperature T

As explained in Section 3.1, growth kinetics and the resulting film morphology can be affected and controlled by varying the deposition rate F (through change in the sputtering power) and the deposition temperature T. This is demonstrated in Figure 6, which plots ρ vs. h_f curves, extracted from in situ SE, during room-temperature (300 K) magnetron-sputter deposition of Ag films on SiO_2/Si substrates (Figure 6a) at two different deposition rates F of 0.15 (red circles) and 0.01 nm/s (black triangles). The data show that an increase of F yields a decrease in h_{cont} (indicated by vertical solid arrows). The ability of the in situ and real time techniques described in Section 2 to establish changes in morphological transition thicknesses (h_{perc} and h_{cont}) as a function of deposition conditions is further illustrated by the $\langle \sigma_f \rangle \times h_f$ and $R_s \times h_f$ vs. h_f curves (Figure 6b,c, respectively). These curves have been recorded during deposition of Ag films on a-C/Si substrates at various values of F and T [61], whereby decrease (increase) of $F(T)$ leads to higher values of h_{perc} and h_{cont}.

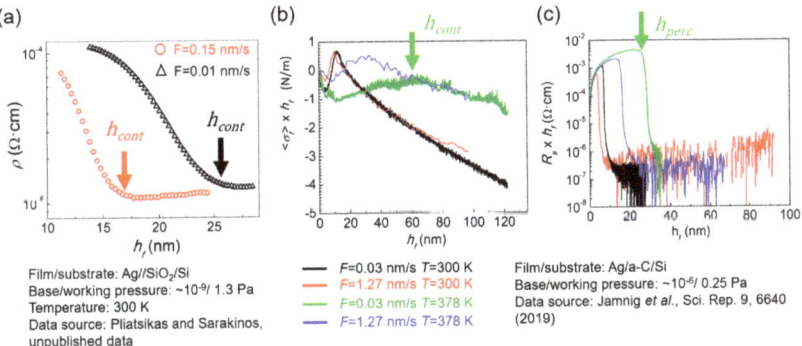

Figure 6. (a) ρ vs. h_f curves, extracted from in situ spectroscopic ellipsometry, during room-temperature (300 K) magnetron-sputter deposition of Ag films on SiO_2/Si substrates at two different deposition rates F of 0.15 (red circles) and 0.01 nm/s (black triangles). The position of h_{cont} on the curves is indicated by solid arrows (unpublished data by Pliatsikas and Sarakinos) (b) $\langle \sigma_f \rangle \times h_f$ vs. h_f curves, extracted from in situ substrate curvature measurements, during deposition of Ag films on a-C/Si substrates at two values of F (0.03 and 1.27 nm/s) and T (300 and 378 K). (c) $R_s \times h_f$ vs. h_f curves, extracted from in situ four-point probe measurements, during deposition of Ag films on a-C/Si substrates at two values of F (0.03 and 1.27 nm/s) and T (300 and 378 K). The positions of h_{cont} and h_{perc} on the curves at $T = 378$ K and $F = 0.03$ nm/s are indicated by solid arrows. Data in (b) and (c) are taken from [61].

To further discuss the effect of deposition conditions on film morphological evolution, we plot in Figure 7 h_{cont} (for Ag layers grown on SiO$_2$ and a-C substrates determined by substrate curvature and ellipsometry measurements as those presented in Figure 6) as a function of F (log–log scale) for various T values. Data are extracted for films deposited by continuous and pulsed magnetron sputtering (filled and hollow symbols, respectively) [61,62,89,91], and by evaporation (half-filled symbols) [99] at different conditions with regards to base and working (i.e., Ar gas in case of sputtering) pressures, as indicated in the respective legends in Figure 7. The first observation (similar to the data in Figure 6) is that, for all conditions and deposition techniques reported in Figure 7, the magnitude of h_{cont} decreases with F, which indicates that 2D growth morphology is favored by larger arrival rates of vapor at the film growth front. As explained in detail in Section 3.1, this can be attributed to either increase of island number density or delay of cluster reshaping during coalescence. In addition, we see that the h_{cont} vs. F data exhibit a power-law dependence (straight lines in log–log scale), in accordance to Equations (7) and (8). Closer inspection of the data reveals that room-temperature (T = 300 K) continuous magnetron-sputter-deposition (black filled symbols) yield negative slopes of 1/7, which is indicative of coalescence-free growth (Equation (8)). Increase of temperature to T = 330 K (blue filled circles) and T = 378 K (red filled squares) results to larger h_{cont} values, i.e., tendency toward 3D morphology is enhanced, as atomic mobility, nucleation, and island reshaping are promoted. Moreover, larger growth temperatures lead to larger negative h_{cont} vs. F slope magnitudes (~1/3), i.e., morphology evolves closer to the coalescence-controlled growth (Equation (7)). By establishing the growth regimes and using h_{cont} data for multiple F and T values, rates of atomic-scale processes that control island nucleation, growth, and coalescence can be estimated [61,91]. Another point of interest in the data plotted in Figure 7, is that for pulsed sputtering (hollow symbols) h_{cont} exhibits a complex scaling dependence on F with h_{cont} vs. F slope changing from ~−1/3 to ~0 with increasing deposition rate. This is a signature of multiple nucleation regimes encountered in pulsed deposition, as explained by Jensen and Niemeyer [110] and Lü et al. [90].

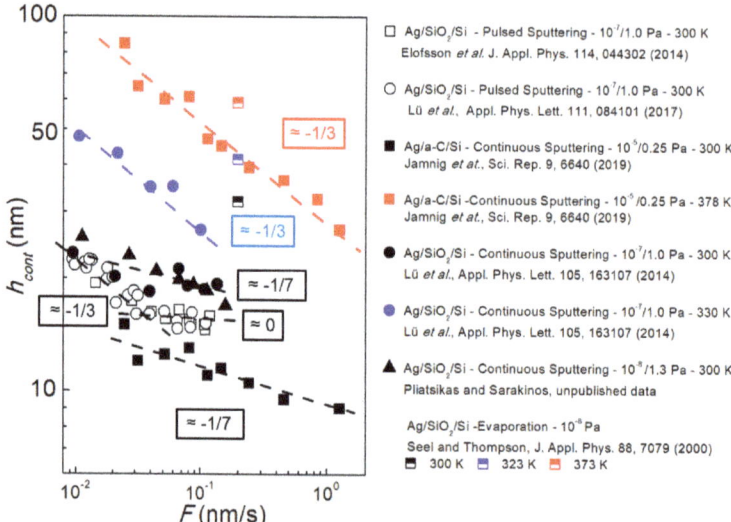

Figure 7. h_{cont} vs. F (log–log scale) at various values of T in the range 300 to 378 K, during growth of Ag on a-C and SiO$_2$ substrates by continuous and pulsed magnetron sputtering (data from [61,62,89,91]) and evaporation (data from [99]). The dashed lines are guides to the eye indicating the h_{cont} vs. F slopes (provided next to each line) for selected set of data. More information on the growth conditions for each data set, including base and working pressure, is provided in the respective legends.

Besides h_{cont}, we also plot in log–log scale in Figure 8 h_{perc} vs. F (h_{perc} determined by four-point-probe measurements as shown in Figure 6) for Ag films grown on a-C using continuous magnetron sputtering, at $T = 300$ K (black filled squares) and $T = 378$ K (red filled squares) [61]. We see that the h_{perc} vs. F line slopes are consistent with those in Figure 7 (h_{cont} vs. F) for both low and high deposition temperatures, while the ratio h_{cont}/h_{perc} takes a value of ~2, irrespective of the growth temperature. This indicates that the morphology set at the early stages of island nucleation and coalescence remains consistent until continuous layer formation. Figure 8 also presents h_{perc} vs. F data obtained during the growth of Ag on SiO_2 by pulsed layer deposition (hollow circles) [111]. This set of data exhibits a negative slope of ~1/7 in the F range 10^{-2} to 10^{-1} nm/s with a tendency for larger slope for $F < 10^{-2}$ nm/s, in qualitative agreement with the h_{cont} vs. F lines for pulsed magnetron sputtering in Figure 7.

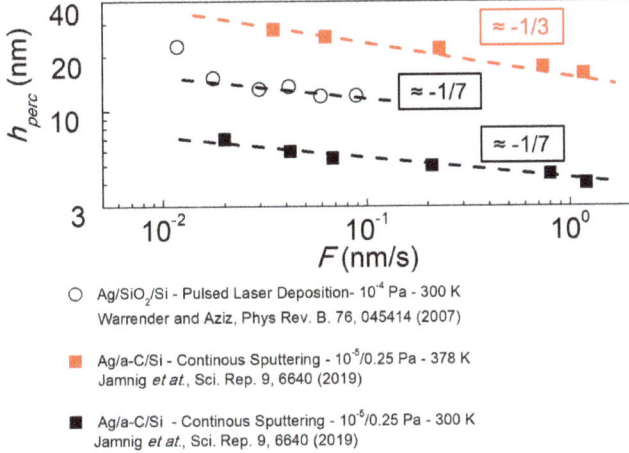

○ Ag/SiO$_2$/Si - Pulsed Laser Deposition- 10^{-4} Pa - 300 K
Warrender and Aziz, Phys Rev. B. 76, 045414 (2007)

■ Ag/a-C/Si - Continous Sputtering - 10^{-8}/0.25 Pa - 378 K
Jamnig et at., Sci. Rep. 9, 6640 (2019)

■ Ag/a-C/Si - Continous Sputtering - 10^{-5}/0.25 Pa - 300 K
Jamnig et at., Sci. Rep. 9, 6640 (2019)

Figure 8. Percolation thickness h_{perc} vs. deposition rate F (log–log scale) at growth temperatures $T = 300$ and 378 K, during the growth of Ag on a-C substrates by continuous sputtering (data from [61]), and pulsed laser deposition (data from [111]). The dashed lines are guides to the eye indicating the h_{perc} vs. F slopes (provided next to each line) for selected set of data. More information on the growth conditions for each data set, including base and working pressure, is provided in the respective legends.

The data presented in Figures 7 and 8 highlight the possibility to fine tune the thickness at which the metallic film becomes electrically conductive or continuous by appropriate control of the vapor flux rate or substrate temperature. For instance, in the case of Ag films, h_{cont} can be varied in a relatively broad range from ~10 nm at room temperature and $F = 1$ nm/s up to ~50 nm at $T = 330$ K and $F = 0.01$ nm/s. This strategy has important implications for practical applications since the control of the early growth stages is decisive in achieving the desired physical attributes which critically depend on the film microstructure.

3.3.3. Other Factors Influencing Film Morphological Evolution

Base pressure: As explained in Section 2.1, data on Ag and Cu were obtained from depositions in vacuum chambers with different base pressures. It can be seen in Figure 7 that h_{cont} is larger for Ag films sputter-deposited in better vacuum conditions (i.e., lower base pressure). This trend is particularly salient at low deposition rates (compare filled triangles vs. filled circles which correspond to base pressures 10^{-8} and 10^{-7} Pa, respectively), and it can be explained by the influence of residual gas contaminants (typically H_2O, O_2, CO) which become adsorbed on the growth surface and act as preferred nucleation sites. This effect has been discussed in the early works by Abermann and Koch [109], who have shown that the tensile peak maximum of the $\langle \sigma_f \rangle \times h_f$ vs. h_f curves recorded for a series of thermally evaporated metal films was shifted to lower values under poorer vacuum conditions. This is most often accompanied by the development of more tensile stress in the continuous film regime, as confirmed in our recent study on sputter-deposited Cu films [112].

Energy of film-forming species: Figure 7 shows that sputter-deposition yields systematically smaller h_{cont} at all temperatures (for a given deposition rate) compared to evaporated films. This can be attributed to the fact that the film growth front during sputtering is exposed to a more energetic (vs. thermalized for evaporation) vapor flux [63,113], which is known to enhance nucleation center density, and thereby promote flat morphology via defect formation and cluster dissociation [22]. Such effects are consistent with earlier reports by Grachev et al. [49] who used in situ and real-time surface differential reflectance spectroscopy and established that the in-plane size of Ag islands formed during sputter-deposition is 3 to 4 times larger compared to Ag islands produced by thermal evaporation. Moreover, for sputter-deposited Cu films, h_{cont} has been shown to decrease from ~16 nm to ~9 nm when decreasing the Ar working pressure from 2 to 0.05 Pa [114], confirming the trend that a larger flux of energetic species to the film growth front (e.g., achieved by decreasing working pressure during sputtering) promotes in-plane island growth and leads to earlier onset of island impingement. This is consistent with the data reported in Figure 7 which also show that the working pressure has influence on the thickness at which Ag films become continuous, with larger h_{cont} values reached at higher Ar working pressure at which energetic sputtered particles experience more collisions in the gas phase, and hence arrive at the substrate with lower kinetic energy and create fewer surface defects. At fixed working pressure, similar trends can be expected by increasing the target-to-substrate distance.

Besides changing the working pressure, the energetic bombardment during growth can be tuned by the application of a negative bias voltage to the substrate. In conventional direct current magnetron discharges, the ionization degree of the sputtered material is very low (1–2%) and working gas ions are the main ionized species in the plasma, so that the application of a bias voltage may result in generation of point defects, larger compressive stress values [113], and decreased charge carrier mobility. However, in high power impulse magnetron sputtering (HiPIMS) discharges, a large fraction of sputtered atoms, up to 50–80%, gets ionized [115], so that applying a bias voltage can be advantageously used to tailor the film microstructure by controlling the energy of film-forming species. Cemin et al. [116] have reported that the thickness at which Cu films are continuous could be increased from h_{cont} ~7 to ~12 nm by increasing the bias from 0 to 130 V, due to enhanced adatom mobility. This results in the formation of Cu films with larger grain size, reduced compressive stress, and higher electrical conductivity [116,117].

Surfactants: The morphology of high-mobility metal films (e.g., Ag) can be further manipulated, for a given set of deposition conditions with regards to F and T, by using minority gaseous (N_2, O_2) or metallic (e.g., Al, Cu, and early transition metals) species to act as wetting agents. Such species can be either deposited as seed layers to change the energy and kinetics at the film/substrate interface and affect nucleation [118], or co-deposited at the film growth front, which modifies the island coalescence dynamics [17,18,79,119–122] as well as stress response [123,124]. In this framework, in situ and real-time growth monitoring methodologies provide insights for selective deployment of wetting agents at key formation stages such that film morphology can be manipulated without compromising other physical properties of the film/substrate system [79,119,120].

3.4. Studies of Discontinuous Ag-Layer Morphology

Section 3.3 has demonstrated the way by which in situ and real-time techniques can be used to determine growth transition thicknesses (h_{perc} and h_{cont}) and their correlation with the dynamics of film morphological evolution. In the present section, we showcase the use of in situ and real-time SE for studying the growth of discontinuous layers and establish the correlation with percolated and continuous films. To this purpose, we use Ag/SiO$_2$ as a model system (Ag is deposited by continuous magnetron sputtering at a base pressure of 10^{-8} Pa and at $F = 0.1$ nm/s) and investigate the effect of deposition temperature ($T = 300$ and 500 K).

Analysis of SE data of electrically conductive layers, using the Drude model and the methodologies explained in Section 2.4, showed that h_{cont} increases from ~26 to >80 nm, when T is increased from 300 to 500 K, i.e., higher temperature promotes 3D (in agreement with the data in Figure 7). Early growth stages of discontinuous layers are investigated by establishing the evolution of LSPR (ellipsometric data analyzed using Lorentz model as explained in Section 2.4) as a function of h_f. The results are shown in Figure 9a, where it is seen that for both deposition temperatures the energy of LSPR decreases (from ~3.3 to ~2.4 eV, i.e., is red-shifted) with increasing h_f. This trend signifies that the mean island size increases and island separation shrinks (i.e., larger fraction of the substrate is covered) as more material is deposited [55,79,119]. Concurrently, we observe that the LSPR vs. h_f slope for $T = 300$ K is steeper than its 500 K counterpart, which means that increase of deposition temperature promotes out-of-plane island growth yielding a lower substrate coverage for a given nominal thickness [79,119]. These differences are confirmed by plan-view scanning electron microscopy (SEM) images of 3 nm thick films at the two growth temperatures (Figure 9b,c), respectively). At $T = 300$ K, the film surface features islands that are partially elongated and interconnected, indicating that incomplete coalescence has started, while the substrate coverage is $\theta = 58\%$. In contrast, at $T = 500$ K, islands are clearly more circular and larger (i.e., island reshaping is completed during coalescence), and $\theta = 44\%$. The qualitative picture and its consistency with the LSPR evolution is further supported by quantitative analysis of the SEM images with regards to the island size distribution and the mean island in-plane aspect ratio (AR). Data (presented in Figure 9d) show that the mean island size MS increases from ~82 to 110 nm² upon temperature increase from 300 to 500 K, while AR decreases from ~1.8 to 1.5 for the same change in temperature (AR = 1 for circular islands).

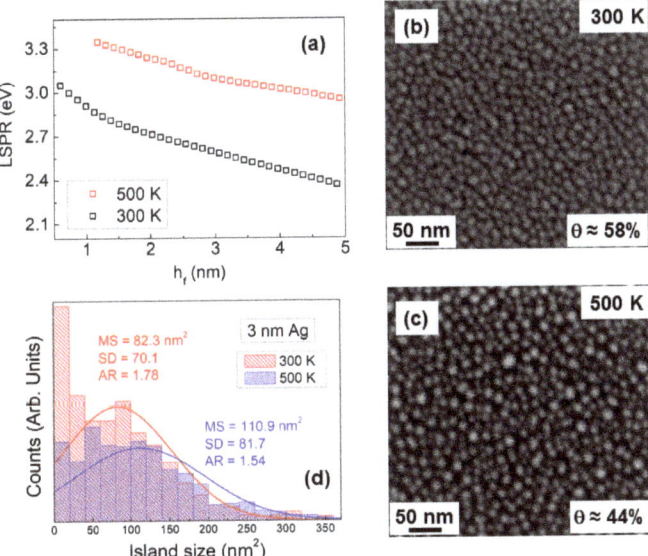

Figure 9. Morphological studies of discontinuous Ag layers deposited by magnetron sputtering on SiO$_2$ substrates (base pressure 10^{-8} Pa; working pressure 1.3 Pa; $F = 0.1$ nm/s) at two deposition temperatures of $T = 300$ and 500 K (unpublished data by Pliatsikas and Sarakinos). (**a**) Evolution of localized surface plasmon resonance (LSPR) vs. nominal film thickness, as determined from in situ and real-time SE. Plan-view SEM images of 3 nm thick films are shown in (**b**) for $T = 300$ K and (**c**) for $T = 500$ K. (**d**) Quantitative analysis of SEM images with regards to island size distribution, mean island size MS, size standard deviation SD, and in-plane aspect ratio AR.

4. Interface Reactivity and Structure Formation

The present section highlights the use of the in situ and real-time diagnostic tools presented in Section 2 for studying more complex scenarios of thin film growth involving chemical reactivity between the deposited metal and the underlying substrate/layer. This is typically the case at the metal/Si interface, where silicide layers usually form by solid-state reaction during annealing [26]. Their formation can also be driven by dynamic processes during film growth, such as segregation or energetic bombardment-induced intermixing. To this purpose, we present and discuss data with regards to growth of bcc transition metals (Mo, Ta, Fe) on Si, which showcase the ability to study amorphous-to crystalline transitions, as well as formation of different crystallographic allotropes during initial film-formation stages. We also demonstrate the potential of combining real-time wafer curvature and resistivity techniques to identify complex interface chemical reactions during growth of Pd on Si and Ge and correlate them to the structure formation during subsequent metal growth.

4.1. Growth of Fe, Mo, and Ta on a-Si

4.1.1. Structure and Phase Formation

We report here the data on real-time monitoring—using MOSS and resistivity diagnostics—of the growth evolution of sputter-deposited Fe, Mo, and Ta films (bcc crystal structure), on Si substrates covered with a 9 nm thick a-Si layer. The a-Si overlayer was sputter-deposited in the same vacuum chamber prior to growth of the metal layer. This methodology provides the same starting growth surface for all experiments and minimizes the influence of oxygen contamination on film growth. Depositions were performed at room temperature (300 K) and at relatively low Ar working pressure (0.24 Pa). Similar deposition rates, as given in Table 2, were employed for all metal layers by controlling the power applied to the respective magnetron sources, such as to rule out possible effects of vapor flux rate F on the observed growth evolution (as discussed in Section 3). At $T = 300$ K, the respective homologous temperatures T_h of the three metals are 0.16 (Fe), 0.10 (Mo), and 0.09 (Ta). Based on this, it is expected that Mo and Ta grow at conditions of limited atomic mobility ($T_h \sim 0.1$), and hence exhibit zone-I type (i.e., underdense) microstructure [63,125,126]. However, atomic mobility is enhanced by energetic bombardment during sputter-deposition at low working pressure [63,113], resulting in fully-dense films, as confirmed by ex situ XRR [127,128].

Table 2. Process parameters and physical quantities extracted from real-time stress and resistivity evolutions during growth of Fe, Mo, and Ta films at $T = 300$ K on a-Si. The Ar working pressure was fixed at 0.24 Pa.

Element	Crystal Structure	Target Power (W)	Deposition Rate F (nm/s)	T_h	Δf (J/m^2)	ξ (nm)	h_{perc} (nm)
Fe	bcc	60	0.06	0.16	2.3 ± 0.1	0.20 ± 0.06	0.30 ± 0.05
Mo	bcc	50	0.05	0.10	3.4 ± 0.1	0.30 ± 0.05	0.27 ± 0.05
Ta	β (A-15)	50	0.05	0.09	3.9 ± 0.1	0.39 ± 0.02	0.14 ± 0.05

Figure 10a,b shows the evolution of $\langle \sigma_f \rangle \times h_f$ and $R_s \times h_f$ vs. h_f during the early growth stages of the three metals, up to $h_f = 10$ nm. Fe and Mo show similar trends of tensile stress build-up and decrease in resistivities upon film thickening. In contrast, the behavior of Ta is distinctly different: the incremental stress becomes compressive for $h_f \geq 2$ nm and the film retains a high electrical resistivity. Ex situ XRD investigations of the crystal structure on thicker films (~60 nm thick) uncover that, while Mo and Fe crystallize in their equilibrium bcc structure, the β-Ta allotrope (tetragonal structure [129,130]) forms in sputter-deposited Ta films [128], as shown in Figure 10c. Hence, the observed differences in the early-stage of stress and resistivity evolutions can be understood as fingerprints of nucleation and growth of different crystallographic phases.

Another striking observation in the curves corresponding to Mo and Fe films presented in Figure 10a,b is the occurrence of transient features which manifest themselves as a concomitant tensile jump and resistivity drop (indicated by arrows). These abrupt variations occur at h_f = 1.9–2.1 nm and h_f = 2.2–2.6 nm for Fe and Mo, respectively, while they are not observed during growth of Ta. Note that the thickness values of these transient features differ by ±0.2 nm between the two in situ techniques since the measurements were not performed simultaneously, but have the same physical origins, as explained in the following. In the case of Ta, no transient feature is observed in the $\langle \sigma_f \rangle \times h_f$ and $R_s \times h_f$ vs. h_f curves, and the β-Ta phase is formed, which is intrinsically more resistive (typical resistivity of 170 μΩ·cm compared to 25–30 μΩ·cm for bcc-Ta [128]) and of lower crystal symmetry. It was proposed that this phase is stabilized by minimization of interface energies [128], although other authors argued that phase formation is governed by the presence of TaO_x interlayer and template (epitaxial) growth [131].

In the case of Mo/a-Si, the transient stress feature observed at around 2.2 nm (a value consistent with literature reports [132]) was shown to correspond to the onset of an amorphous-to-crystalline (a–c) phase transition [133,134], which was unambiguously established recently by coupling in situ MOSS and XRD measurements [135]. The structure formation pathway in the case of Mo deposition on a-Si can be summarized as follows: (i) Mo atoms react with Si surface atoms, forming an amorphous silicide interfacial layer over a thickness of ~2 nm [132]; (ii) at a critical thickness of 2.2 ± 0.1 nm, the amorphous layer crystallizes into bcc Mo. The a–c transition results in a volume contraction (film densification) and structural ordering, which are reflected by the formation of tensile stress and decrease in film resistivity (Figure 10a,b), respectively. Due to the observed similarities among the real-time curves of Fe and Mo, we argue that the structure formation process during growth of Fe on a-Si obeys the same scenario (as for Mo). Steiner et al. [136] reported the formation of an interfacial amorphous layer during growth of Fe film on B for $h_f \le 3$ nm. Interestingly, a drop in electrical resistance was evidenced at h_f~3 nm, likely imputable to crystallization, although the authors for [136] did not explicitly draw this conclusion.

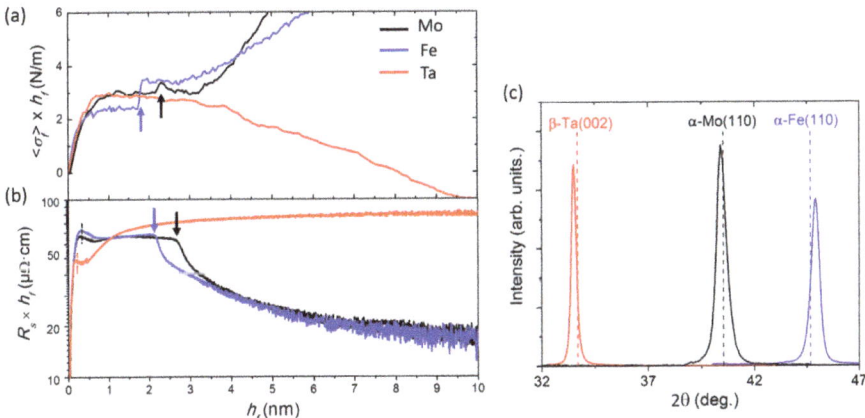

Figure 10. Real-time evolution of (a) $\langle \sigma_f \rangle \times h_f$ and (b) $R_s \times h_f$ curves vs. film thickness h_f during the early growth stages of Fe, Mo, and Ta films on a-Si at T = 300 K (deposition conditions are indicated in Table 2). The arrows in (a) and (b) indicate the onset of film crystallization for Fe and Mo. (c) XRD patterns recorded in Bragg–Brentano configuration from ~60 nm thick Fe, Mo, and Ta films using Cu Kα radiation (λ = 0.15418 nm). MOSS data are taken from [105,128,134].

4.1.2. Early Growth Morphology and Interface Stress

In the present section, we focus on the incipient growth stages ($h_f < 2$ nm) of Mo, Ta, and Fe films, as evidenced by the data presented in Figure 10a,b. The evolution of the $R_s \times h_f$ vs. h_f curves shows an initial drop below 0.5 nm (indicated by dashed lines in (Figure 10b), which can be attributed to film percolation. Compared to the h_{perc} values reported in Section 3 for Ag, Cu, and Pd films deposited on weakly-interacting substrates, the percolation thickness found for Fe, Mo, and Ta are extremely small, around 0.3–0.4 nm (see Table 2), which corresponds to the deposition of ~2 monolayers of film materials. This clearly shows that Mo, Ta, and Fe exhibit a pronounced 2D (i.e., wetting) growth morphology on a-Si, further supported by AFM imaging of layers with thicknesses in the range of 5 nm [71].

Concomitantly, the $\langle \sigma_f \rangle \times h_f$ curves (plotted for $0 \leq h_f \leq 2$ nm in Figure 11a) show the formation of tensile stress, an opposite behavior to what is observed in Figure 3 for metal films on weakly-interacting substrates. The tensile stress levels off until a plateau is reached, which can be understood by the change in surface/interface stress Δf due to the progressive coverage of the a-Si surface (surface energy γ^{Si}) by a metal (Me) deposit ($\gamma^{Me} > \gamma^{Si}$). The quantities f and γ are of the same order of magnitude, and for an isotropic solid surface are related through the expression $f = \gamma + \frac{\partial \gamma}{\partial \varepsilon}$, where ε is the elastic strain within the interface plane [137,138]. Then, the asymptotic behavior of the $\langle \sigma_f \rangle \times h_f$ vs. h_f curves in Figure 11a can be fitted by an exponential interaction term, following the approach proposed by Müller and Thomas [138], according to

$$\langle \sigma_f \rangle \times h_f = \Delta f \left(1 - e^{-h_f/\xi}\right) \tag{9}$$

where $\Delta f = f^{Me} + f^{int} - f^{Si}$ is the change in surface/interface stress during deposition of metal (Me) on a-Si, and ξ is a characteristic interaction length that accounts for the finite extension of the interface stress contribution. Δf and ξ are extracted by fitting Equation (9) to the MOSS data and values for Fe, Mo, and Ta are reported in Table 2. An example of the fit is shown in Figure 11a for the case of Fe. It is seen that the ξ values are of the same order of magnitude as h_{perc} determined from resistivity measurements. Moreover, ξ increases with increasing atomic number of the metal, which may be attributed to larger interface spread (intermixing) due to more intense energetic bombardment. In the case of the Mo/a-Si interface, Reinink et al. [132] confirmed the formation of an interfacial MoSi$_x$ alloy from in situ low energy ion scattering analysis, and concluded that the tensile stress was due to compound formation, rather than a change in surface stress. Nevertheless, as seen in Figure 11b, the derived Δf values are found to scale with the change in surface energy, $\Delta \gamma = \gamma^{Me} - \gamma^{Si}$, for different metals with either bcc (present work and from [71]) or hcp structure (unpublished data by Abadias). This suggests that the early growth stages of metals on a-Si are dominated by interfacial effects.

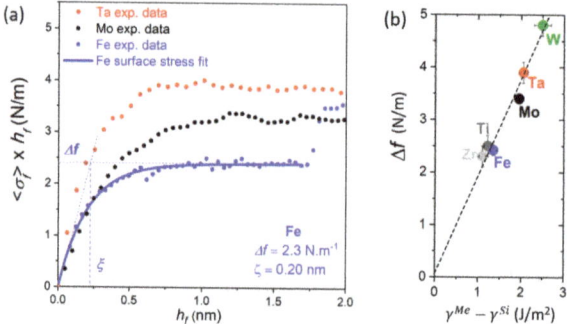

Figure 11. (a) Enlarged view of the stress-thickness evolution at the Me/a-Si interface during growth of Fe, Mo, and Ta on a-Si. An analytical model, Equation (9), is used to fit experimental data (see solid line

for the case of Fe) to extract Δf and ξ. (b) Plot of Δf vs. $\Delta\gamma$ for bcc Fe, Mo, Ta and W (data obtained from present work and Refs. [71,105]) and hcp Ti and Zr (unpublished data by Abadias) deposited on a-Si. and. Surface energies γ^{Me} and γ^{Si} are taken from [139].

4.2. Effect of Interface Reactivity on Morphological Evolution

In order to demonstrate the importance of interface reactivity on film nucleation and growth, we compare here the morphological evolution of Pd layers deposited on different substrates: SiO_2, a-Si and a-Ge. Figure 12a,b show the $\langle\sigma_f\rangle \times h_f$ and $R_s \times h_f$ vs. h_f real-time data collected for Pd films deposited at $T = 300$ K and $T = 355$ K, respectively. The blue curves correspond to the growth of Pd on SiO_2, as already described in Section 3.3.1, and serve as reference. Overall, we notice that the curves corresponding to growth on a-Si and a-Ge are more complex than those for growth on SiO_2, especially for $h_f < 5$ nm, where the following features are observed: (i) a tensile stress is clearly formed from the beginning of growth, followed by a transient peak at ~2 nm (clearly visible on a-Ge at $T = 300$ K and a-Si at $T = 355$ K, see arrow, and smeared out in the case of a-Si at $T = 300$ K); (ii) the $R_s \times h_f$ vs. h_f curve exhibits two consecutive drops at $h_f \sim 2$ nm (marked by arrow) and below 4 nm (indicated by dashed lines). The evolution of the Pd/a-Si and Pd/a-Ge curves in Figure 12a for $h_f > 5$ nm is reminiscent to what is observed for the growth of Pd on SiO_2 beyond h_{cont}, i.e., a compressive stress regime is established for $h_f > h_{cont}$, accompanied by a continuous decrease of the film resistivity. It can be noted, however, that the $R_s \times h_f$ values for Pd films deposited on a-Si and a-Ge are higher than that of the Pd film deposited on SiO_2, while the magnitude of the incremental stress remains similar. For the Pd/a-Si system, the overall stress and resistivity evolutions are similar between $T = 300$ K and $T = 355$ K (Figure 12b). Differences in the stress development were found at higher temperatures due to higher interface reactivity [105], but will not be discussed here.

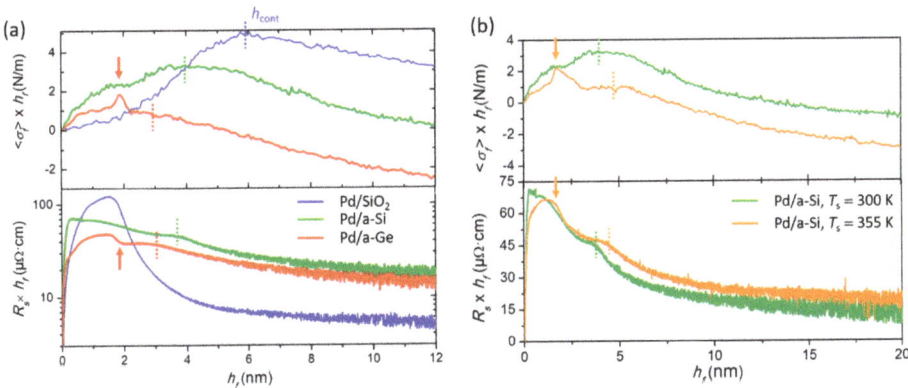

Figure 12. Real-time evolution of $\langle\sigma_f\rangle \times h_f$ and $R_s \times h_f$ curves vs. h_f during sputter-deposition of Pd films on (a) SiO_2, a-Si and a-Ge at $T = 300$ K and (b) a-Si at $T = 300$ and 355 K. The curves exhibit distinct features at critical transition thickness, marked by arrows and vertical dashed lines (for interpretation, see text).

The $\langle\sigma_f\rangle \times h_f$ vs. h_f evolution at the Pd/a-Si and Pd/a-Ge interface exhibit the same asymptotic tensile behavior as observed during growth of Fe, Mo, and Ta on a-Si (see Figure 11a), suggesting the formation of an intermixed layer. It is well known that Pd atoms react with Si to form a silicide compound with composition close to Pd_2Si, usually obtained after thermal annealing [140–143]. The silicide formation is accompanied by the build-up of stress, being either tensile or compressive depending on the diffusion process [26,141,144]. Only few reports have studied the growth of Pd/a-Si at room temperature [140,143]. Ex situ XRD and TEM characterization performed on the Pd/a-Si system confirm the formation of a silicide compound Pd_2Si [105]. Figure 13a,c shows plan-view TEM micrographs obtained from a 9 nm thick Pd film deposited on a-Si. Crystalline regions are visible, corresponding either to fcc Pd or hex-Pd$_2$Si, further supported by electron diffraction pattern (Figure 13b).

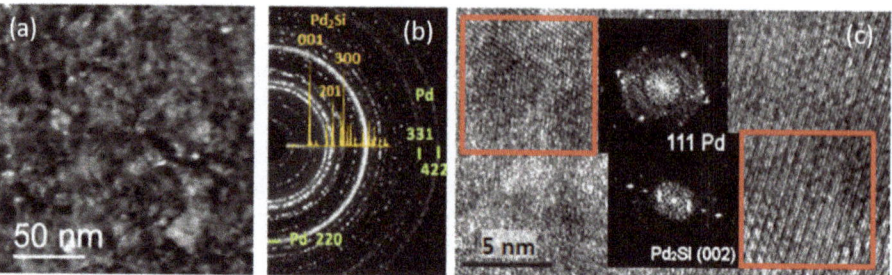

Figure 13. Plan-view TEM investigation of a 9 nm thick Pd film deposited on a-Si. (**a**) STEM-HAADF image revealing the grain contrast in the thin film. (**b**) selected area electron diffraction pattern showing the coexistence of polycrystalline Pd$_2$Si and 111 fiber-textured Pd grains. (**c**) High resolution image; Fourier transform on selected grains are consistent either with fcc Pd or hexagonal Pd$_2$Si lattice planes.

As first proposed in [105], the simultaneous occurrence of a tensile stress peak and resistivity drop at $h_f \sim 2$ nm during growth of Pd on a-Si was ascribed to the crystallization of a Pd$_2$Si interfacial compound. This scenario was confirmed and more precisely described in a recent work that combined simultaneously MOSS and XRD measurements using synchrotron facilities (beamline SIXS at SOLEIL) [145]. The authors could reveal that a nanocrystalline or amorphous Pd$_2$Si layer is first formed, which suddenly crystallizes giving rise to the observed tensile peak and resistivity drop at 2.3 nm. Then, the Pd layer nucleates while the [111] textured Pd$_2$Si layer continues to grow up to a thickness of 3.7 nm. The second drop in resistivity observed in Figure 12 around 3–4 nm could be the fingerprint of the formation of the elemental and more conductive Pd layer. As shown in Figure 12b, the tensile peak is better resolved and sharper at $T = 355$ K due to thermally-enhanced interfacial reaction. The similarities between the stress and resistivity curves recorded on a-Ge at $T = 300$ K and a-Si at $T = 355$ K point towards a higher reactivity between Pd and Ge atoms.

5. Conclusions and Outlook

The results presented herein demonstrate that non-invasive and easy to implement optical and electrical probes can be used as efficient in situ and real-time diagnostics to characterize the early growth stages of polycrystalline metallic films on semiconducting or insulating substrates. Examples are provided for a wide variety of metal/substrate systems for which different wetting behaviors and morphological evolutions are clearly revealed, primarily dictated by differences in surface mobility and interaction of the metal adatoms with the substrate layer. In the case of elemental metals deposited on weakly-interacting substrates (SiO$_2$, a-C) and growing in 3D fashion (like Ag, Cu, or Pd), wafer curvature, spectroscopic ellipsometry, and electrical resistivity measurements are complementary in monitoring the relevant film formation stages, and determining morphological transition thicknesses corresponding to percolation and formation of a continuous layer. Through a systematic study of the dependence of these thicknesses with deposition rate F and substrate temperature T, the scaling behavior of these quantities has been established, from which two kinetic growth regimes could be identified. These regimes correspond to a coalescence-free and coalescence-controlled evolution of the island shapes.

More generally, the impact of deposition process parameters (working pressure, bias voltage, target-to-substrate distance) on film morphology can be directly evaluated by means of these real-time diagnostics, and establish the following empirical rules for controlling film morphological evolution: (i) for transparent conductors, it is necessary to form a percolated film at the lowest possible thickness, which can be achieved by decreasing substrate temperature, increasing deposition rate, or decreasing working pressure; (ii) opposite variations of the latter parameters decrease wetting and yield nanogranular films/supported nanoparticles. Application of a bias voltage may have an opposite effect on the onset of film percolation and formation of a continuous layer, depending on the type of sputtering discharge (DC vs. Hipims).

Our in situ and real-time nanoscale monitoring can be readily applied for studying the growth of other metals like Au, Pt, or Ti in the prospect of fabricating smart coatings capable to enhance the absorption of light in photovoltaics [146] or manage the light in novel dielectric–metal–dielectric transparent electrodes [147,148]. The developed methodology can also be extended to investigate the impact of alloying elements or minority species acting as surfactant on the film morphology and stress response. It also offers the possibility to explore more complex systems like transition metal nitrides, thin film metallic glasses or high-entropy alloys. Concurrently, it would be also very interesting to employ these real-time diagnostics for a better understanding of the mechanisms controlling solid-state dewetting of continuous layers [149–151], which is a promising route to fabricate nanogranular films.

The potential of coupling in situ and real-time stress and resistivity measurements is also leveraged to unveil structural transitions and interfacial reactions during growth of ultrathin transition metal (Fe, Mo, and Ta) and Pd layers on amorphous Si or Ge. These systems are characterized by a 2D growth morphology and the formation of an intermixed layer at the metal/Si (or Ge) interface over a thickness of ~1 nm. For Mo/Si and Fe/Si systems, the interfacial silicide compound is amorphous, and the metal layer crystallizes into its equilibrium bcc structure at a critical thickness of ~2 nm, as detected in real-time from the stress and resistivity transient features. Ta films grow instead in a tetragonal structure, with no visible change in the electrical and stress properties.

The overall results reviewed in the present article demonstrate that in situ, real-time optical and electrical techniques are powerful tools to monitor thin film growth at the nanoscale and gain insights on the correlation among atomic-scale mechanisms and resulting film morphology. From the view point of applications, these diagnostics could further contribute to optimize the growth of functional nanoscale layers and smart coating technologies relying on the integration of plasmonic devices and transparent electrodes on substrates beyond Si, such as glass, oxides, Van-der-Waals 2D crystals, or polymeric/flexible substrates [17,146–148,152–154]. To this end, their implementation may involve some technological improvement/modification of the actual set-up (for instance the curvature of a transparent substrate from laser-based deflectometry can be measured from the substrate backside coated with a reflective layer or by using other types of stress-sensors), as well as refinement of the optical modelling of ellipsometric data (to account for interband transitions in some metallic films, or optical anisotropy of some polymeric substrates). Combining these approaches with in situ X-ray-based techniques will enable us to get a detailed understanding of the structure formation and film morphology, which opens the path to engineering the film morphological and physical attributes by appropriate control of the process parameters.

Author Contributions: Conceptualization, J.C., K.S. and G.A.; methodology, J.C., A.J., K.S. and G.A.; validation, K.S. and G.A.; formal analysis, J.C., A.J. and N.P.; investigation, J.C., A.J., A.M., N.P. and C.F.; data curation, J.C., A.J. and N.P.; writing—original draft preparation, J.C.; writing—review and editing, K.S. and G.A.; supervision, K.S. and G.A.; funding acquisition, K.S. and G.A. All authors have read and agreed to the published version of the manuscript.

Funding: A.J. and G.A. acknowledge the financial support of the French Government program "Investissements d'Avenir" (LABEX INTERACTIFS, reference ANR-11-LABX-0017-01). This work also pertains to the French Government program "Investissements d'Avenir" (EUR INTREE, reference ANR-18-EURE-0010). K.S. acknowledges financial support from Linköping University ("LiU Career Contract, Dnr-LiU-2015-01510, 2015-2020") and the Swedish research council (contract VR-2015-04630). A.J. and K.S. acknowledge financial support from the ÅForsk foundation (contracts ÅF 19-137 and ÅF 19-746). K.S. and N.P. acknowledge financial support from the Olle Engkvist foundation (contract SOEB 190-312) and the Wenner-Gren foundations (contracts UPD2018-0071 and UPD2019-0007).

Acknowledgments: The authors wish to acknowledge the technical and scientific support of Philippe Guérin (PPrime Institute) for the in situ resistivity set-up, and his responsibility for the maintenance of the deposition chamber.

Conflicts of Interest: The authors declare no conflict of interest.

References

1. Campbell, C.T. Metal films and particles on oxide surfaces: Structural, electronic and chemisorptive properties. *J. Chem. Soc. Faraday Trans.* **1996**, *92*, 1435–1445. [CrossRef]
2. Mueller, T.; Xia, F.; Avouris, P. Graphene photodetectors for high-speed optical communications. *Nat. Photonics* **2010**, *4*, 297–301. [CrossRef]
3. Gong, C.; Huang, C.; Miller, J.; Cheng, L.; Hao, Y.; Cobden, D.; Kim, J.; Ruoff, R.S.; Wallace, R.M.; Cho, K.; et al. Metal contacts on p hysical vapor deposited monolayer MoS2. *ACS Nano* **2013**, *7*, 11350–11357. [CrossRef] [PubMed]
4. Liu, X.; Han, Y.; Evans, J.W.; Engstfeld, A.K.; Behm, R.J.; Tringides, M.C.; Hupalo, M.; Lin, H.Q.; Huang, L.; Ho, K.M.; et al. Growth morphology and properties of metals on graphene. *Prog. Surf. Sci.* **2015**, *90*, 397–443. [CrossRef]
5. Echtermeyer, T.J.; Milana, S.; Sassi, U.; Eiden, A.; Wu, M.; Lidorikis, E.; Ferrari, A.C. Surface Plasmon Polariton Graphene Photodetectors. *Nano Lett.* **2016**, *16*, 8–20. [CrossRef] [PubMed]
6. Zilberberg, K.; Riedl, T. Metal-nanostructures-a modern and powerful platform to create transparent electrodes for thin-film photovoltaics. *J. Mater. Chem. A* **2016**, *4*, 14481–14508. [CrossRef]
7. Yun, J. Ultrathin Metal films for Transparent Electrodes of Flexible Optoelectronic Devices. *Adv. Funct. Mater.* **2017**, *27*, 1606641. [CrossRef]
8. Xu, Y.; Hsieh, C.-Y.; Wu, L.; Ang, L.K. Two-dimensional transition metal dichalcogenides mediated long range surface plasmon resonance biosensors. *J. Phys. D Appl. Phys.* **2019**, *52*, 65101. [CrossRef]
9. Willets, K.A.; Van Duyne, R.P. Localized surface plasmon resonance spectroscopy and sensing. *Annu. Rev. Phys. Chem.* **2007**, *58*, 267–297. [CrossRef]
10. Stockman, M.I.; Kneipp, K.; Bozhevolnyi, S.I.; Saha, S.; Dutta, A.; Ndukaife, J.; Kinsey, N.; Reddy, H.; Guler, U.; Shalaev, V.M.; et al. Roadmap on plasmonics. *J. Opt. UK* **2018**, *20*, 43001. [CrossRef]
11. Patsalas, P.; Kalfagiannis, N.; Kassavetis, S.; Abadias, G.; Bellas, D.V.; Lekka, C.; Lidorikis, E. Conductive nitrides: Growth principles, optical and electronic properties, and their perspectives in photonics and plasmonics. *Mater. Sci. Eng. R Rep.* **2018**, *123*, 1–55. [CrossRef]
12. Zhang, Y.; He, S.; Guo, W.; Hu, Y.; Huang, J.; Mulcahy, J.R.; Wei, W.D. Surface-Plasmon-Driven Hot Electron Photochemistry. *Chem. Rev.* **2018**, *118*, 2927–2954. [CrossRef] [PubMed]
13. Goetz, S.; Bauch, M.; Dimopoulos, T.; Trassl, S. Ultrathin sputter-deposited plasmonic silver nanostructures. *Nanoscale Adv.* **2020**, *2*, 869–877. [CrossRef]
14. Achanta, V.G. Surface waves at metal-dielectric interfaces: Material science perspective. *Rev. Phys.* **2020**, *5*, 100041. [CrossRef]
15. Liu, L.; Corma, A. Metal Catalysts for Heterogeneous Catalysis: From Single Atoms to Nanoclusters and Nanoparticles. *Chem. Rev.* **2018**, *118*, 4981–5079. [CrossRef]
16. Kato, K.; Omoto, H.; Tomioka, T.; Takamatsu, A. Visible and near infrared light absorbance of Ag thin films deposited on ZnO underlayers by magnetron sputtering. *Sol. Energy Mater. Sol. Cells* **2011**, *95*, 2352–2356. [CrossRef]
17. Wang, W.; Song, M.; Bae, T.-S.; Park, Y.H.; Kang, Y.-C.; Lee, S.-G.; Kim, S.-Y.; Kim, D.H.; Lee, S.; Min, G.; et al. Transparent Ultrathin Oxygen-Doped Silver Electrodes for Flexible Organic Solar Cells. *Adv. Funct. Mater.* **2014**, *24*, 1551–1561. [CrossRef]
18. Zhao, G.; Wang, W.; Bae, T.S.; Lee, S.S.G.; Mun, C.W.; Lee, S.S.G.; Yu, H.; Lee, G.H.; Song, M.; Yun, J. Stable ultrathin partially oxidized copper film electrode for highly efficient flexible solar cells. *Nat. Commun.* **2015**, *6*, 8830. [CrossRef]
19. Venables, J.A.; Spiller, G.D.T.; Hanbucken, M. Nucleation and growth of thin films. *Rep. Prog. Phys.* **1984**, *47*, 399–459. [CrossRef]
20. Barna, P.B.; Adamik, M. Fundamental structure forming phenomena of polycrystalline films and the structure zone models. *Thin Solid Films* **1998**, *317*, 27–33. [CrossRef]
21. Petrov, I.; Barna, P.B.; Hultman, L.; Greene, J.E. Microstructural evolution during film growth. *J. Vac. Sci. Technol. A* **2003**, *21*, S117–S128. [CrossRef]
22. Michely, T.; Krug, J. *Islands, Mounds and Atoms*; Springer: Berlin/Heidelberg, Germany, 2004.

23. Thompson, C.V. Structure evolution during processing of polycrystalline films. *Annu. Rev. Mater. Sci.* **2000**, *30*, 159–190. [CrossRef]
24. Greene, J.E. Thin Film Nucleation, Growth, and Microstructural Evolution: An Atomic Scale View. *Handb. Depos. Technol. Film. Coat.* **2010**, 554–620. [CrossRef]
25. Martin, P.M. (Ed.) *Handbook of Deposition Technologies for Films and Coatings*, 3rd ed.; William Andrew: Amsterdam, The Netherlands, 2010; ISBN 9780815520313.
26. d'Heurle, F.M.; Gas, P. Kinetics of formation of silicides: A review. *J. Mater. Res.* **1986**, *1*, 205–221. [CrossRef]
27. Laurila, T.; Molarius, J. Reactive Phase Formation in Thin Film Metal/Metal and Metal/Silicon Diffusion Couples. *Crit. Rev. Solid State Mater. Sci.* **2003**, *28*, 185–230. [CrossRef]
28. Creuze, J.; Berthier, F.; Tétot, R.; Legrand, B. Phase transition induced by superficial segregation: The respective role of the size mismatch and of the chemistry. *Surf. Sci.* **2001**, *491*, 1–16. [CrossRef]
29. Marks, L.D.; Peng, L. Nanoparticle shape, thermodynamics and kinetics. *J. Phys. Condens. Matter* **2016**, *28*, 53001. [CrossRef] [PubMed]
30. Marsault, M.; Sitja, G.; Henry, C.R. Regular arrays of Pd and PdAu clusters on ultrathin alumina films for reactivity studies. *Phys. Chem. Chem. Phys.* **2014**, *16*, 26458–26466. [CrossRef]
31. Rost, M.J.; Quist, D.A.; Frenken, J.W.M. Grains, Growth, and Grooving. *Phys. Rev. Lett.* **2003**, *91*, 26101. [CrossRef]
32. Rost, M.J.; Crama, L.; Schakel, P.; Van Tol, E.; Van Velzen-Williams, G.B.E.M.; Overgauw, C.F.; Ter Horst, H.; Dekker, H.; Okhuijsen, B.; Seynen, M.; et al. Scanning probe microscopes go video rate and beyond. *Rev. Sci. Instrum.* **2005**, *76*, 53710. [CrossRef]
33. Bauer, E. Low energy electron microscopy. *Rep. Prog. Phys.* **1994**, *57*, 895–938. [CrossRef]
34. Tromp, R.M. Low-energy electron microscopy. *IBM J. Res. Dev.* **2000**, *44*, 503–516. [CrossRef]
35. Cheynis, F.; Leroy, F.; Ranguis, A.; Detailleur, B.; Bindzi, P.; Veit, C.; Bon, W.; Müller, P. Combining low-energy electron microscopy and scanning probe microscopy techniques for surface science: Development of a novel sample-holder. *Rev. Sci. Instrum.* **2014**, *85*, 43705. [CrossRef]
36. Devloo-Casier, K.; Ludwig, K.F.; Detavernier, C.; Dendooven, J. In situ synchrotron based x-ray techniques as monitoring tools for atomic layer deposition. *J. Vac. Sci. Technol. A Vac. Surf. Films* **2014**, *32*, 10801. [CrossRef]
37. Dendooven, J.; Solano, E.; Minjauw, M.M.; Van De Kerckhove, K.; Coati, A.; Fonda, E.; Portale, G.; Garreau, Y.; Detavernier, C. Mobile setup for synchrotron based in situ characterization during thermal and plasma-enhanced atomic layer deposition. *Rev. Sci. Instrum.* **2016**, *87*, 113905. [CrossRef]
38. Renaud, G.; Lazzari, R.; Revenant, C.; Barbier, A.; Noblet, M.; Ulrich, O.; Leroy, F.; Jupille, J.; Borensztein, Y.; Henry, C.R.; et al. Real-time monitoring of growing nanoparticles. *Science* **2003**, *300*, 1416–1419. [CrossRef]
39. Renaud, G.; Lazzari, R.; Leroy, F. Probing surface and interface morphology with Grazing Incidence Small Angle X-ray Scattering. *Surf. Sci. Rep.* **2009**, *64*, 255–380. [CrossRef]
40. Hodas, M.; Siffalovic, P.; Jergel, M.; Pelletta, M.; Halahovets, Y.; Vegso, K.; Kotlar, M.; Majkova, E. Kinetics of copper growth on graphene revealed by time-resolved small-angle x-ray scattering. *Phys. Rev. B* **2017**, *95*, 35424. [CrossRef]
41. Schwartzkopf, M.; Santoro, G.; Brett, C.J.; Rothkirch, A.; Polonskyi, O.; Hinz, A.; Metwalli, E.; Yao, Y.; Strunskus, T.; Faupel, F.; et al. Real-Time Monitoring of Morphology and Optical Properties during Sputter Deposition for Tailoring Metal-Polymer Interfaces. *ACS Appl. Mater. Interfaces* **2015**, *7*, 13547–13556. [CrossRef]
42. Schwartzkopf, M.; Roth, S.V. Investigating polymer–metal interfaces by grazing incidence small-angle x-ray scattering from gradients to real-time studies. *Nanomaterials* **2016**, *6*, 239. [CrossRef]
43. Dann, E.K.; Gibson, E.K.; Catlow, R.A.; Collier, P.; Eralp Erden, T.; Gianolio, D.; Hardacre, C.; Kroner, A.; Raj, A.; Goguet, A.; et al. Combined in Situ XAFS/DRIFTS Studies of the Evolution of Nanoparticle Structures from Molecular Precursors. *Chem. Mater.* **2017**, *29*, 7515–7523. [CrossRef]
44. Schuisky, M.; Elam, J.W.; George, S.M. In situ resistivity measurements during the atomic layer deposition of ZnO and W thin films. *Appl. Phys. Lett.* **2002**, *81*, 180–182. [CrossRef]
45. Arnalds, U.B.; Agustsson, J.S.; Ingason, A.S.; Eriksson, A.K.; Gylfason, K.B.; Gudmundsson, J.T.; Olafsson, S. A magnetron sputtering system for the preparation of patterned thin films and in situ thin film electrical resistance measurements. *Rev. Sci. Instrum.* **2007**, *78*, 103901. [CrossRef]

46. Agustsson, J.S.; Arnalds, U.B.; Ingason, A.S.; Gylfason, K.B.; Johnsen, K.; Olafsson, S.; Gudmundsson, J.T. Growth, coalescence, and electrical resistivity of thin Pt films grown by dc magnetron sputtering on SiO_2. *Appl. Surf. Sci.* **2008**, *254*, 7356–7360. [CrossRef]
47. Colin, J.J.; Diot, Y.; Guerin, P.; Lamongie, B.; Berneau, F.; Michel, A.; Jaouen, C.; Abadias, G. A load-lock compatible system for in situ electrical resistivity measurements during thin film growth. *Rev. Sci. Instrum.* **2016**, *87*, 23902. [CrossRef]
48. Simonot, L.; Babonneau, D.; Camelio, S.; Lantiat, D.; Guérin, P.; Lamongie, B.; Antad, V. In situ optical spectroscopy during deposition of Ag:Si3N4 nanocomposite films by magnetron sputtering. *Thin Solid Films* **2010**, *518*, 2637–2643. [CrossRef]
49. Grachev, S.; De Grazia, M.; Barthel, E.; Søndergård, E.; Lazzari, R. Real-time monitoring of nanoparticle film growth at high deposition rate with optical spectroscopy of plasmon resonances. *J. Phys. D Appl. Phys.* **2013**, *46*, 375305. [CrossRef]
50. Renaud, G.; Ducruet, M.; Ulrich, O.; Lazzari, R. Apparatus for real time in situ quantitative studies of growing nanoparticles by grazing incidence small angle X-ray scattering and surface differential reflectance spectroscopy. *Nucl. Instrum. Methods Phys. Res. Sect. B Beam Interact. Mater. Atoms* **2004**, *222*, 667–680. [CrossRef]
51. Lazzari, R.; Jupille, J. Quantitative analysis of nanoparticle growth through plasmonics. *Nanotechnology* **2011**, *22*, 445703. [CrossRef]
52. An, I.; Li, Y.M.; Nguyen, H.V.; Collins, R.W. Spectroscopic ellipsometry on the millisecond time scale for real-time investigations of thin-film and surface phenomena. *Rev. Sci. Instrum.* **1992**, *63*, 3842–3848. [CrossRef]
53. Collins, R.W.; An, I.; Nguyen, H.V.; Lu, Y. Real time spectroscopic ellipsometry for characterization of nucleation, growth, and optical functions of thin films. *Thin Solid Films* **1993**, *233*, 244–252. [CrossRef]
54. Oates, T.W.H.; McKenzie, D.R.; Bilek, M.M.M. Percolation threshold in ultrathin titanium films determined by in situ spectroscopic ellipsometry. *Phys. Rev. B* **2004**, *70*, 195406. [CrossRef]
55. Oates, T.W.H.; Mücklich, A. Evolution of plasmon resonances during plasma deposition of silver nanoparticles. *Nanotechnology* **2005**, *16*, 2606–2611. [CrossRef]
56. Patsalas, P.; Logothetidis, S. Interface properties and structural evolution of TiN/Si and TiN/GaN heterostructures. *J. Appl. Phys.* **2003**, *93*, 989–998. [CrossRef]
57. Bulíř, J.; Novotný, M.; Lančok, J.; Fekete, L.; Drahokoupil, J.; Musil, J. Nucleation of ultrathin silver layer by magnetron sputtering in Ar/N2 plasma. *Surf. Coat. Technol.* **2013**, *228*, S86–S90. [CrossRef]
58. Amassian, A.; Desjardins, P.; Martinu, L. Study of TiO2 film growth mechanisms in low-pressure plasma by in situ real-time spectroscopic ellipsometry. *Thin Solid Films* **2004**, *447*, 40–45. [CrossRef]
59. Walker, J.D.; Khatri, H.; Ranjan, V.; Li, J.; Collins, R.W.; Marsillac, S. Electronic and structural properties of molybdenum thin films as determined by real-time spectroscopic ellipsometry. *Appl. Phys. Lett.* **2009**, *94*, 141908. [CrossRef]
60. Wu, P.C.; Losurdo, M.; Kim, T.-H.; Choi, S.; Bruno, G.; Brown, A.S. In situ spectroscopic ellipsometry to monitor surface plasmon resonant group-III metals deposited by molecular beam epitaxy. *J. Vac. Sci. Technol. B Microelectron. Nanom. Struct.* **2007**, *25*, 1019–1023. [CrossRef]
61. Jamnig, A.; Sangiovanni, D.G.; Abadias, G.; Sarakinos, K. Atomic-scale diffusion rates during growth of thin metal films on weakly-interacting substrates. *Sci. Rep.* **2019**, *9*, 6640. [CrossRef]
62. Elofsson, V.; Lü, B.; Magnfält, D.; Münger, E.P.; Sarakinos, K. Unravelling the physical mechanisms that determine microstructural evolution of ultrathin Volmer-Weber films. *J. Appl. Phys.* **2014**, *116*, 44302. [CrossRef]
63. Thornton, J.A. The microstructure of sputter-deposited coatings. *J. Vac. Sci. Technol. A Vac. Surf. Films* **1986**, *4*, 3059–3065. [CrossRef]
64. Thornton, J.A.; Hoffman, D.W. Stress-related effects in thin-films. *Thin Solid Films* **1989**, *171*, 5–31. [CrossRef]
65. Depla, D.; Braeckman, B.R. Quantitative correlation between intrinsic stress and microstructure of thin films. *Thin Solid Films* **2016**, *604*, 90–93. [CrossRef]
66. Gambino, J.P.; Colgan, E.G. Silicides and ohmic contacts. *Mater. Chem. Phys.* **1998**, *52*, 99–146. [CrossRef]
67. Abadias, G.; Chason, E.; Keckes, J.; Sebastiani, M.; Thompson, G.B.; Barthel, E.; Doll, G.L.; Murray, C.E.; Stoessel, C.H.; Martinu, L. Review Article: Stress in thin films and coatings: Current status, challenges, and prospects. *J. Vac. Sci. Technol. A* **2018**, *36*, 20801. [CrossRef]

68. Fluri, A.; Schneider, C.W.; Pergolesi, D. In situ stress measurements of metal oxide thin films. In *Metal Oxide-Based Thin Film Structures*; Pryds, N., Esposito, V., Eds.; Elsevier Inc.: Amsterdam, The Neatherlands, 2018; pp. 109–132.
69. Floro, J.A.; Chason, E.; Lee, S.R. Real time measurement of epilayer strain using a simplified wafer curvature technique. *Mater. Res. Soc. Symp. Proc.* **1996**, *406*, 491–496. [CrossRef]
70. Chason, E.; Guduru, P.R. Tutorial: Understanding residual stress in polycrystalline thin films through real-time measurements and physical models. *J. Appl. Phys.* **2016**, *119*, 191101. [CrossRef]
71. Abadias, G.; Fillon, A.; Colin, J.J.; Michel, A.; Jaouen, C. Real-time stress evolution during early growth stages of sputter-deposited metal films: Influence of adatom mobility. *Vacuum* **2014**, *100*, 36–40. [CrossRef]
72. Le Priol, A.; Simonot, L.; Abadias, G.; Guérin, P.; Renault, P.-O.O.; Le Bourhis, E. Real-time curvature and optical spectroscopy monitoring of magnetron-sputtered WTi alloy thin films. *Surf. Coat. Technol.* **2013**, *237*, 112–117. [CrossRef]
73. Abadias, G.; Simonot, L.; Colin, J.J.; Michel, A.; Camelio, S.; Babonneau, D. Volmer-Weber growth stages of polycrystalline metal films probed by in situ and real-time optical diagnostics. *Appl. Phys. Lett.* **2015**, *107*, 183105. [CrossRef]
74. Azzam, R.M.A.; Bashara, N.M. *Ellipsometry and Polarised Light*; North Holland: Amsterdam, The Netherlands, 1977.
75. Herzinger, C.M.; Johs, B.; McGahan, W.A.; Woollam, J.A.; Paulson, W. Ellipsometric determination of optical constants for silicon and thermally grown silicon dioxide via a multi-sample, multi-wavelength, multi-angle investigation. *J. Appl. Phys.* **1998**, *83*, 3323–3336. [CrossRef]
76. Wooten, F. *Optical Properties of Solids*; Academic Press: New York, NY, USA, 1972.
77. Oates, T.W.H.; Wormeester, H.; Arwin, H. Characterization of plasmonic effects in thin films and metamaterials using spectroscopic ellipsometry. *Prog. Surf. Sci.* **2011**, *86*, 328–376. [CrossRef]
78. Furgeaud, C. Effets Cinétique et Chimique lors des Premiers Stades de Croissance de Films Minces méTalliques: Compréhension Multi-échelle par une Approche Expérimentale et Modélisation Numérique. Ph.D. Thesis, University of Poitiers, Poitiers, France, 2019.
79. Jamnig, A.; Pliatsikas, N.; Konpan, M.; Lu, J.; Kehagias, T.; Kotanidis, A.N.; Kalfagiannis, N.; Bellas, D.V.; Lidorikis, E.; Kovac, J.; et al. 3D-to-2D Morphology Manipulation of Sputter-Deposited Nanoscale Silver Films on Weakly Interacting Substrates via Selective Nitrogen Deployment for Multifunctional Metal Contacts. *ACS Appl. Nano Mater.* **2020**, *3*, 4728–4738. [CrossRef]
80. Zinsmeister, G. A contribution to Frenkel's theory of condensation. *Vacuum* **1966**, *16*, 529–535. [CrossRef]
81. Kryukov, Y.A.; Amar, J.G. Scaling of the island density and island-size distribution in irreversible submonolayer growth of three-dimensional islands. *Phys. Rev. B* **2010**, *81*, 165435. [CrossRef]
82. Nichols, F.A.; Mullins, W.W. Morphological Changes of a Surface of Revolution due to Capillarity-Induced Surface Diffusion. *J. Appl. Phys.* **1965**, *36*, 1826–1835. [CrossRef]
83. Nichols, F.A. Coalescence of Two Spheres by Surface Diffusion. *J. Appl. Phys.* **1966**, *37*, 2805–2808. [CrossRef]
84. Jeffers, G.; Dubson, M.A.; Duxbury, P.M. Island-to-percolation transition during growth of metal films. *J. Appl. Phys.* **1994**, *75*, 5016–5020. [CrossRef]
85. Family, F.; Meakin, P. Scaling of the Droplet-Size Distribution in Vapor-Deposited Thin Films. *Phys. Rev. Lett.* **1988**, *61*, 428–431. [CrossRef]
86. Jensen, P. Growth of nanostructures by cluster deposition: Experiments and simple models. *Rev. Mod. Phys.* **1999**, *71*, 1695–1735. [CrossRef]
87. Carrey, J.; Maurice, J.-L. Transition from droplet growth to percolation: Monte Carlo simulations and an analytical model. *Phys. Rev. B* **2001**, *63*, 245408. [CrossRef]
88. Carrey, J.; Maurice, J.-L. Scaling laws near percolation during three-dimensional cluster growth: A Monte Carlo study. *Phys. Rev. B* **2002**, *65*, 205401. [CrossRef]
89. Lü, B.; Elofsson, V.; Münger, E.P.; Sarakinos, K. Dynamic competition between island growth and coalescence in metal-on-insulator deposition. *Appl. Phys. Lett.* **2014**, *105*, 163107. [CrossRef]
90. Lü, B.; Münger, E.P.; Sarakinos, K. Coalescence-controlled and coalescence-free growth regimes during deposition of pulsed metal vapor fluxes on insulating surfaces. *J. Appl. Phys.* **2015**, *117*, 134304. [CrossRef]
91. Lü, B.; Souqui, L.; Elofsson, V.; Sarakinos, K. Scaling of elongation transition thickness during thin-film growth on weakly interacting substrates. *Appl. Phys. Lett.* **2017**, *111*, 84101. [CrossRef]
92. Maaroof, A.I.; Evans, B.L. Onset of electrical conduction in Pt and Ni films. *J. Appl. Phys.* **1994**, *76*, 1047–1054. [CrossRef]

93. Rycroft, I.M.; Evans, B.L. The in situ characterization of metal film resistance during deposition. *Thin Solid Films* **1996**, *290*, 283–288. [CrossRef]
94. Hérault, Q. Vers la Compréhension de la Croissance des Couches Minces d'argent par pulvérisation, à la lumière de Mesures Operando. Ph.D. Thesis, Sorbonne Université, Paris, France, 2019.
95. Floro, J.A.; Chason, E.; Cammarata, R.C.; Srolovitz, D.J. Physical origins of intrinsic stresses in Volmer-Weber thin films. *MRS Bull.* **2002**, *27*, 19–25. [CrossRef]
96. Cammarata, R.C.; Trimble, T.M.; Srolovitz, D.J. Surface stress model for intrinsic stresses in thin films. *J. Mater. Res.* **2000**, *15*, 2468–2474. [CrossRef]
97. Hoffman, R.W. Stresses in thin films: The relevance of grain boundaries and impurities. *Thin Solid Films* **1976**, *34*, 185–190. [CrossRef]
98. Nix, W.D.; Clemens, B.M. Crystallite coalescence: A mechanism for intrinsic tensile stresses in thin films. *J. Mater. Res.* **1999**, *14*, 3467–3473. [CrossRef]
99. Seel, S.C.; Thompson, C.V.; Hearne, S.J.; Floro, J.A. Tensile stress evolution during deposition of Volmer–Weber thin films. *J. Appl. Phys.* **2000**, *88*, 7079–7088. [CrossRef]
100. Chason, E.; Sheldon, B.W.; Freund, L.B.; Floro, J.A.; Hearne, S.J. Origin of compressive residual stress in polycrystalline thin films. *Phys. Rev. Lett.* **2002**, *88*, 156103. [CrossRef]
101. Koch, R.; Hu, D.; Das, A.K. Compressive stress in polycrystalline Volmer-Weber films. *Phys. Rev. Lett.* **2005**, *94*, 146101. [CrossRef]
102. Chason, E. A kinetic analysis of residual stress evolution in polycrystalline thin films. *Thin Solid Films* **2012**, *526*, 1–14. [CrossRef]
103. Yu, H.Z.; Thompson, C.V. Correlation of shape changes of grain surfaces and reversible stress evolution during interruptions of polycrystalline film growth. *Appl. Phys. Lett.* **2014**, *104*, 141913. [CrossRef]
104. Vasco, E.; Polop, C. Intrinsic Compressive Stress in Polycrystalline Films is Localized at Edges of the Grain Boundaries. *Phys. Rev. Lett.* **2017**, *119*, 256102. [CrossRef] [PubMed]
105. Colin, J.J. Potentialités des Techniques de Caractérisation In-Situ et en temps réel pour sonder, Comprendre et Contrôler les Processus de Nucléation-croissance Durant le dépôt de Films Minces Métalliques. Ph.D. Thesis, University of Poitiers, Poitiers, France, 2015.
106. Lü, B.; Almyras, G.A.; Gervilla, V.; Greene, J.E.; Sarakinos, K. Formation and morphological evolution of self-similar 3D nanostructures on weakly interacting substrates. *Phys. Rev. Mater.* **2018**, *2*, 63401. [CrossRef]
107. Gervilla, V.; Almyras, G.A.; Thunström, F.; Greene, J.E.; Sarakinos, K. Dynamics of 3D-island growth on weakly-interacting substrates. *Appl. Surf. Sci.* **2019**, *488*, 383–390. [CrossRef]
108. Gervilla, V.; Almyras, G.A.; Lü, B.; Sarakinos, K. Coalescence dynamics of 3D islands on weakly-interacting substrates. *Sci. Rep.* **2020**, *10*, 2031. [CrossRef]
109. Abermann, R.; Koch, R. In situ study of thin film growth by internal stress measurement under ultrahigh vacuum conditions: Silver and copper under the influence of oxygen. *Thin Solid Films* **1986**, *142*, 65–76. [CrossRef]
110. Jensen, P.; Niemeyer, B. The effect of a modulated flux on the growth of thin films. *Surf. Sci.* **1997**, *384*, L823–L827. [CrossRef]
111. Warrender, J.M.; Aziz, M.J. Effect of deposition rate on morphology evolution of metal-on-insulator films grown by pulsed laser deposition. *Phys. Rev. B* **2007**, *76*, 45414. [CrossRef]
112. Jamnig, A.; Pliatsikas, N.; Sarakinos, K.; Abadias, G. The effect of kinetics on intrinsic stress generation and evolution in sputter-deposited films at conditions of high atomic mobility. *J. Appl. Phys.* **2020**, *127*, 45302. [CrossRef]
113. Windischmann, H. Intrinsic stress in sputter-deposited thin films. *Crit. Rev. Solid State Mater. Sci.* **1992**, *17*, 547–596. [CrossRef]
114. Pletea, M.; Brückner, W.; Wendrock, H.; Kaltofen, R. Stress evolution during and after sputter deposition of Cu thin films onto Si (100) substrates under various sputtering pressures. *J. Appl. Phys.* **2005**, *97*, 54908. [CrossRef]
115. Sarakinos, K.; Alami, J.; Konstantinidis, S. High power pulsed magnetron sputtering: A review on scientific and engineering state of the art. *Surf. Coat. Technol.* **2010**, *204*, 1661–1684. [CrossRef]
116. Cemin, F.; Abadias, G.; Minea, T.; Furgeaud, C.; Brisset, F.; Solas, D.; Lundin, D. Benefits of energetic ion bombardment for tailoring stress and microstructural evolution during growth of Cu thin films. *Acta Mater.* **2017**, *141*, 120–130. [CrossRef]

117. Cemin, F.; Lundin, D.; Cammilleri, D.; Maroutian, T.; Lecoeur, P.; Minea, T. Low electrical resistivity in thin and ultrathin copper layers grown by high power impulse magnetron sputtering. *J. Vac. Sci. Technol. A Vac. Surf. Films* **2016**, *34*, 51506. [CrossRef]
118. Anders, A.; Byon, E.; Kim, D.H.; Fukuda, K.; Lim, S.H.N. Smoothing of ultrathin silver films by transition metal seeding. *Solid State Commun.* **2006**, *140*, 225–229. [CrossRef]
119. Pliatsikas, N.; Jamnig, A.; Konpan, M.; Delimitis, A.; Abadias, G.; Sarakinos, K. Manipulation of thin silver film growth on weakly interacting silicon dioxide substrates using oxygen as a surfactant Manipulation of thin silver film growth on weakly interacting silicon dioxide substrates using oxygen as a surfactant. *J. Vac. Sci. Technol. A* **2020**, *38*, 43406. [CrossRef]
120. Jamnig, A.; Pliatsikas, N.; Abadias, G.; Sarakinos, K. On the effect of copper as wetting agent during growth of thin silver films on silicon dioxide substrates. *Appl. Surf. Sci.* **2020**, *538*, 148056. [CrossRef]
121. Zhao, G.; Jeong, E.; Choi, E.A.; Yu, S.M.; Bae, J.S.; Lee, S.G.; Han, S.Z.; Lee, G.H.; Yun, J. Strategy for improving Ag wetting on oxides: Coalescence dynamics versus nucleation density. *Appl. Surf. Sci.* **2020**, *510*, 145515. [CrossRef]
122. Yun, J.; Chung, H.S.; Lee, S.G.; Bae, J.S.; Hong, T.E.; Takahashi, K.; Yu, S.M.; Park, J.; Guo, Q.; Lee, G.H.; et al. An unexpected surfactant role of immiscible nitrogen in the structural development of silver nanoparticles: An experimental and numerical investigation. *Nanoscale* **2020**, *12*, 1749–1758. [CrossRef] [PubMed]
123. Kaub, T.M.; Felfer, P.; Cairney, J.M.; Thompson, G.B. Influence of Ni Solute segregation on the intrinsic growth stresses in Cu(Ni) thin films. *Scr. Mater.* **2016**, *113*, 131–134. [CrossRef]
124. Kaub, T.; Anthony, R.; Thompson, G.B. Intrinsic stress response of low and high mobility solute additions to Cu thin films. *J. Appl. Phys.* **2017**, *122*, 225302. [CrossRef]
125. Movchan, B.A.; Demchishin, W.V. Study of the Structure and Properties of Thick Vacuum Condensates of Nickel, Titanium, Tungsten, Aluminium Oxide and Zirconium Dioxide. *Phys. Met. Met.* **1969**, *28*, 83–90.
126. Mahieu, S.; Ghekiere, P.; Depla, D.; De Gryse, R. Biaxial alignment in sputter deposited thin films. *Thin Solid Films* **2006**, *515*, 1229–1249. [CrossRef]
127. Belliard, L.; Huynh, A.; Perrin, B.; Michel, A.; Abadias, G.; Jaouen, C. Elastic properties and phonon generation in Mo/Si superlattices. *Phys. Rev. B Condens. Matter Mater. Phys.* **2009**, *80*, 155424. [CrossRef]
128. Colin, J.J.; Abadias, G.; Michel, A.; Jaouen, C. On the origin of the metastable β-Ta phase stabilization in tantalum sputtered thin films. *Acta Mater.* **2017**, *126*, 481–493. [CrossRef]
129. Arakcheeva, A.; Chapuis, G.; Grinevitch, V. The self-hosting structure of β-Ta. *Acta Crystallogr. Sect. B Struct. Crystallogr. Cryst. Chem.* **2002**, *58*, 1–7. [CrossRef]
130. Abadias, G.; Colin, J.J.; Tingaud, D.; Djemia, P.; Belliard, L.; Tromas, C. Elastic properties of α- and β-tantalum thin films. *Thin Solid Films* **2019**, *688*, 137403. [CrossRef]
131. Ellis, E.A.I.; Chmielus, M.; Baker, S.P. Effect of sputter pressure on Ta thin films: Beta phase formation, texture, and stresses. *Acta Mater.* **2018**, *150*, 317–326. [CrossRef]
132. Reinink, J.; Zameshin, A.; Van De Kruijs, R.W.E.; Bijkerk, F. In-situ studies of silicide formation during growth of molybdenum-silicon interfaces. *J. Appl. Phys.* **2019**, *126*, 135304. [CrossRef]
133. Bajt, S.; Stearns, D.G.; Kearney, P.A. Investigation of the amorphous-to-crystalline transition in Mo/Si multilayers. *J. Appl. Phys.* **2001**, *90*, 1017–1025. [CrossRef]
134. Fillon, A.; Abadias, G.; Michel, A.; Jaouen, C.; Villechaise, P. Influence of Phase Transformation on Stress Evolution during Growth of Metal Thin Films on Silicon. *Phys. Rev. Lett.* **2010**, *104*, 96101. [CrossRef]
135. Krause, B.; Abadias, G.; Michel, A.; Wochner, P.; Ibrahimkutty, S.; Baumbach, T. Direct Observation of the Thickness-Induced Crystallization and Stress Build-Up during Sputter-Deposition of Nanoscale Silicide Films. *ACS Appl. Mater. Interfaces* **2016**, *8*, 34888–34895. [CrossRef]
136. Steiner, R.; Boyen, H.G.; Krieger, M.; Plettl, A.; Widmayer, P.; Ziemann, P.; Banhart, F.; Kilper, R.; Oelhafen, P. Interface reactions in [Fe/B]n multilayers: A way to tune from crystalline/amorphous layer sequences to homogeneous amorphous FexB100-x films. *Appl. Phys. A Mater. Sci. Process.* **2003**, *76*, 5–13. [CrossRef]
137. Cammarata, R.C. Surface and interface stress effects on the growth of thin films. *Prog. Surf. Sci.* **1994**, *46*, 1–38. [CrossRef]
138. Müller, P.; Thomas, O. Asymptotic behaviour of stress establishment in thin films. *Surf. Sci.* **2000**, *465*, L764–L770. [CrossRef]
139. Jiang, Q.; Lu, H.M. Size dependent interface energy and its application. *Surf. Sci. Rep.* **2008**, *63*, 427–464. [CrossRef]

140. Edelman, F.; Cytermann, C.; Brener, R.; Eizenberg, M.; Weil, R.; Beyer, W. Interfacial reactions in the Pd/a-Si/c-Si system. *J. Appl. Phys.* **1992**, *71*, 289–295. [CrossRef]
141. Richard, M.I.; Fouet, J.; Guichet, C.; Mocuta, C.; Thomas, O. Exploring Pd-Si(001) and Pd-Si(111) thin-film reactions by simultaneous synchrotron X-ray diffraction and substrate curvature measurements. *Thin Solid Films* **2013**, *530*, 100–104. [CrossRef]
142. Richard, M.I.; Fouet, J.; Texier, M.; Mocuta, C.; Guichet, C.; Thomas, O. Continuous and Collective Grain Rotation in Nanoscale Thin Films during Silicidation. *Phys. Rev. Lett.* **2015**, *115*, 266101. [CrossRef]
143. Molina-Ruiz, M.; Ferrando-Villalba, P.; Rodríguez-Tinoco, C.; Garcia, G.; Rodríguez-Viejo, J.; Peral, I.; Lopeandía, A.F. Simultaneous nanocalorimetry and fast XRD measurements to study the silicide formation in Pd/a-Si bilayers. *J. Synchrotron Radiat.* **2015**, *22*, 717–722. [CrossRef]
144. d'Heurle, F.M.; Thomas, O. Stresses during silicide formation: A review. *Defect Diffus. Forum* **1996**, *129*, 137–150. [CrossRef]
145. Krause, B.; Abadias, G.; Furgeaud, C.; Michel, A.; Resta, A.; Coati, A.; Garreau, Y.; Vlad, A.; Hauschild, D.; Baumbach, T. Interfacial Silicide Formation and Stress Evolution during Sputter-Deposition of Ultrathin Pd Layers on a-Si. *ACS Appl. Mater. Interfaces* **2019**, *11*, 39315–39323. [CrossRef]
146. De Melo, C.; Jullien, M.; Battie, Y.; En Naciri, A.; Ghanbaja, J.; Montaigne, F.; Pierson, J.F.; Rigoni, F.; Almqvist, N.; Vomiero, A.; et al. Tunable Localized Surface Plasmon Resonance and Broadband Visible Photoresponse of Cu Nanoparticles/ZnO Surfaces. *ACS Appl. Mater. Interfaces* **2018**, *10*, 40958–40965. [CrossRef]
147. Kim, S.; Lee, J.-L. Design of dielectric/metal/dielectric transparent electrodes for flexible electronics. *J. Photonics Energy* **2012**, *2*, 21215. [CrossRef]
148. Kim, H.-J.; Seo, K.-W.; Kim, H.-K.; Noh, Y.-J.; Na, S.-I. Ag-Pd-Cu alloy inserted transparent indium tin oxide electrodes for organic solar cells. *J. Vac. Sci. Technol. A Vac. Surf. Films* **2014**, *32*, 51507. [CrossRef]
149. Thompson, C.V. Solid-state dewetting of thin films. *Annu. Rev. Mater. Res.* **2012**, *42*, 399–434. [CrossRef]
150. Jacquet, P.; Podor, R.; Ravaux, J.; Lautru, J.; Teisseire, J.; Gozhyk, I.; Jupille, J.; Lazzari, R. On the solid-state dewetting of polycrystalline thin films: Capillary versus grain growth approach. *Acta Mater.* **2018**, *143*, 281–290. [CrossRef]
151. Leroy, F.; Cheynis, F.; Almadori, Y.; Curiotto, S.; Trautmann, M.; Barbé, J.C.; Müller, P. How to control solid state dewetting: A short review. *Surf. Sci. Rep.* **2016**, *71*, 391–409. [CrossRef]
152. Zhao, G.; Kim, S.M.; Lee, S.-G.; Bae, T.-S.; Mun, C.; Lee, S.; Yu, H.; Lee, G.-H.; Lee, H.-S.; Song, M.; et al. Bendable Solar Cells from Stable, Flexible, and Transparent Conducting Electrodes Fabricated Using a Nitrogen-Doped Ultrathin Copper Film. *Adv. Funct. Mater.* **2016**, *26*, 4180–4191. [CrossRef]
153. Zhao, G.; Shen, W.; Jeong, E.; Lee, S.G.; Yu, S.M.; Bae, T.S.; Lee, G.H.; Han, S.Z.; Tang, J.; Choi, E.A.; et al. Ultrathin Silver Film Electrodes with Ultralow Optical and Electrical Losses for Flexible Organic Photovoltaics. *ACS Appl. Mater. Interfaces* **2018**, *10*, 27510–27520. [CrossRef]
154. De Melo, C.; Jullien, M.; Battie, Y.; En Naciri, A.; Ghanbaja, J.; Montaigne, F.; Pierson, J.F.; Rigoni, F.; Almqvist, N.; Vomiero, A.; et al. Semi-Transparent p-Cu2O/n-ZnO Nanoscale-film heterojunctions for photodetection and photovoltaic applications. *ACS Appl. Nano Mater.* **2019**, *2*, 4358–4366. [CrossRef]

Publisher's Note: MDPI stays neutral with regard to jurisdictional claims in published maps and institutional affiliations.

© 2020 by the authors. Licensee MDPI, Basel, Switzerland. This article is an open access article distributed under the terms and conditions of the Creative Commons Attribution (CC BY) license (http://creativecommons.org/licenses/by/4.0/).

Review

Manipulation and Applications of Hotspots in Nanostructured Surfaces and Thin Films

Xiaoyu Zhao [1,2], Jiahong Wen [3,*], Aonan Zhu [4], Mingyu Cheng [4], Qi Zhu [4], Xiaolong Zhang [4,*], Yaxin Wang [1] and Yongjun Zhang [1,*]

1. Innovative Center for Advanced Materials (ICAM), School of Material and Environmental Engineering, Hangzhou Dianzi University, Hangzhou 310018, China; zhaoxy@hdu.edu.cn (X.Z.); wangyaxin1010@126.com (Y.W.)
2. National Laboratory of Solid State Microstructures, Nanjing University, Nanjing 210093, China
3. The College of Electronics and Information, Hangzhou Dianzi University, Hangzhou 310018, China
4. Key Laboratory of Functional Materials Physics and Chemistry, Ministry of Education, College of Physics, Jilin Normal University, Changchun 130103, China; aonanzhu@126.com (A.Z.); chengmingyu0531@163.com (M.C.); qizhu12300@163.com (Q.Z.)
* Correspondence: wenjiahong@hdu.edu.cn (J.W.); zhangxiaolong@jlnu.edu.cn (X.Z.); yjzhang@hdu.edu.cn (Y.Z.)

Received: 19 July 2020; Accepted: 6 August 2020; Published: 26 August 2020

Abstract: The synthesis of nanostructured surfaces and thin films has potential applications in the field of plasmonics, including plasmon sensors, plasmon-enhanced molecular spectroscopy (PEMS), plasmon-mediated chemical reactions (PMCRs), and so on. In this article, we review various nanostructured surfaces and thin films obtained by the combination of nanosphere lithography (NSL) and physical vapor deposition. Plasmonic nanostructured surfaces and thin films can be fabricated by controlling the deposition process, etching time, transfer, fabrication routes, and their combination steps, which manipulate the formation, distribution, and evolution of hotspots. Based on these hotspots, PEMS and PMCRs can be achieved. This is especially significant for the early diagnosis of hepatocellular carcinoma (HCC) based on surface-enhanced Raman scattering (SERS) and controlling the growth locations of Ag nanoparticles (AgNPs) in nanostructured surfaces and thin films, which is expected to enhance the optical and sensing performance.

Keywords: nanostructured surfaces and thin films; physical vapor deposition; nanosphere lithography; manipulation and applications of hotspots

---

1. Introduction

The synthesis of nanostructured surfaces and thin films using physical vapor deposition, such as pulsed laser deposition, magnetron sputtering, thermal evaporation, e-beam evaporation, among others, plays a key role in the development of a variety of applications in nanoplasmonics, nanoscale photovoltaic devices, nanogenerators, flexible or nanobiological sensors, and so on [1–8]. For devices based on nanostructured surfaces and thin films, diverse high-fidelity geometry is important for the performance of the devices in practical applications. Reliable artificial nanopatterned surfaces and thin films are fabricated by advanced lithographic methods, including electron-beam lithography (EBL), photolithography, soft lithography, nanosphere lithography (NSL), and many others [9–12]. For example, the combined processes of EBL, metal deposition, and liftoff are utilized to obtain patterned metallic structures on a scale of tens of nanometers to submillimeter [12]. However, the multiple wet processes in EBL-based methodology is extremely time-consuming and may introduce additional contaminations on the nanostructured surfaces, which may have non-negligible effects on the quality of nanostructured surfaces and thus degrade the performance of the devices. More importantly, it significantly hinders the direct applicability of the devices, especially in the field of nanostructure-based plasmonics, generators, sensors, and so on. Though many efforts have been made to prepare nanostructured surfaces and thin films by EBL-based methodology, challenges remain in the manipulation of hotspots that are usually observed in the sub-10 nm metallic nanogaps, where the energy is localized to subwavelength dimensions due to the design of the nanostructured surfaces and thin films.

Compared to the EBL-based methodology, the NSL-based approach has attracted much attention. This method uses self-assembled polystyrene (PS) colloid sphere arrays as ordered templates/masks to manipulate hotspots in nanostructured surfaces and thin films. The method is rapid, simple, low-cost, practical, and produces no pollution [13]. Various nanostructured surfaces and thin films can be achieved by the combination of NSL and physical vapor deposition, such as periodic nanocaps [14–18], nanotriangles [19–22], nanobowls [23–25], nanorings [26–28], nanopillars [29,30], nanocones [31–33], and other complex nanostructured surfaces and thin films, including nanohoneycomb, bridged knobby units, nanoparticle cluster-in-bowl arrays, and so on [2,25,34–40]. These architectural designs of nanostructured surfaces and thin films can be obtained by controlling a series of deposition processes (the deposition time, angle, distance, and so on), PS colloid sphere etching, transfer, and their combination steps, which manipulate the formation, distribution, and evolution of hotspots and have significant implications in broad applications [41–55]. Based on these manipulation of hotspots in nanostructure-based surfaces and thin films, plasmon-enhanced molecular spectroscopy (PEMS) and plasmon-mediated chemical reactions (PMCRs) can be controlled, which is expected to enhance the optical and sensing performance.

In this article, the design and synthesis of nanostructured surfaces and thin films with various hybridization of nanoshape arrays are discussed in detail, including large-area periodic nanohoneycomb, nanocap star, nanoring nanoparticle, bridged knobby units, and three-dimensional (3D) nanopillar cap arrays. The formation, distribution, and evolution of hotspots in these nanostructured surfaces and thin films are controlled, which has potential applications in PEMS and PMCRs. Hopefully, this article will inspire more ingenious designs of nanostructured surfaces and thin films using the NSL technique to manipulate hotspots, which is expected to enhance the optical and sensing performance.

2. Experimental Section

2.1. NSL Technique and Physical Vapor Deposition Technique

The NSL technique originates from self-assembled monolayer nanospheres being used as a mask to achieve large-area surface-patterned nanostructures, also known as "natural lithography" [49]. The nanosphere particles are arranged in an ordered array by spin coating, Langmuir–Blodgett technique, electrophoretic deposition, micropropulsive injection (MPI) method, and so on [56–58]. Then, defect-free PS colloid sphere arrays from single layer (SL) to multilayer (ML) are fabricated, which greatly extends the application of "natural lithography" and is called NSL [21,27,50,59]. Due to developments over the past several decades, the NSL technique has been recognized as an effective way to fabricate large-scale ordered nanostructured surfaces and thin films with various nanopatterns [34,42,60]. Normally, NSL have three main processes. First, SL or ML closely packed PS colloid sphere arrays are prepared by the self-assembly method [27,28,56–58,61]. Briefly, PS colloid sphere particles are dispersed in alcoholic solution, which is slowly dripped on the surface of a Si wafer. Then, the Si wafer is slowly immersed into the container filled with deionized water. At the interface between the PS colloid sphere particles and deionized water, the PS colloid spheres start to form an unordered monolayer. After that, the monolayer is driven into a highly ordered array by interactions including van der Waals forces, steric repulsions, and Coulombic repulsions [57]. Then, the highly ordered PS colloid sphere array is picked up by the hydrophilic property substrate. The detailed preparation process of large-scale ordered PS colloid sphere arrays is reported in our previous works [9,13,17,18]. Then, the size, nanogaps, and surface morphology of the as-prepared PS colloid sphere arrays is modified by physical and chemical method, which serves as a template [29,30,37–39,53]. Finally, various materials, including metal, metal oxides, polymers, and so on, are deposited on the ordered PS colloid sphere array templates by physical vapor deposition technique (for example, pulsed laser deposition, magnetron sputtering, thermal evaporation, e-beam evaporation), which make the nanostructured surfaces and thin films more functional.

2.2. Design of Nanostructured Surfaces and Thin Films and Manipulation of Hotspots

Based on the combination of the as-prepared PS colloid sphere array templates and physical vapor deposition, various nanostructured surfaces and thin films can be designed by adjusting the fabrication routes and deposition parameters. Under this strategy, PS colloid sphere arrays with a fixed diameter (e.g., 100, 200, 500, 1000 nm) are selected as templates, and the etching, transfer, rotation, co-sputtering, glancing angle sputtering, single- or multilayer deposition, or their combinations is performed to control the shape of the PS colloid spheres, nanogaps between neighboring PS colloid spheres, the thickness of films, the surface morphology of films, and other parameters of the complex structures. The variety of nanostructured surfaces and thin films can be expanded, obtaining a set of novel nanopatterned arrays by the corresponding control strategy. Several types of novel nanopatterned arrays, including nanohoneycomb, nanocap star, nanoring nanoparticle, bridged knobby units, nanopillar cap, and other hybrid nanostructure arrays, are described in detail in the next section.

At the nanostructured surfaces of some noble metals (e.g., Au, Ag, etc.), the coherent oscillations of the conduction electrons can be driven by light, which will cause localized surface plasmon resonances (LSPRs). Usually, the electromagnetic (EM) field is enormously enhanced in the region of LSPRs, which is called "hotspots". As is well known, these hotspots are usually observed in the sharp tips or corners, sub-10 nm metallic nanogaps, and so on. Moreover, the formation, distribution, and evolution of hotspots are very sensitive to the material, composition, and surrounding dielectric environment of the nanostructures. Therefore, the manipulation of hotspots can be achieved by adjusting the aspects mentioned above. To confirm the formation, distribution, and evolution of hotspots, finite-difference time-domain (FDTD) software is used to simulate the EM field of the nanostructured surfaces and thin films, which has guiding significance on the design of nanostructured surfaces and thin films and the manipulation of hotspots.

2.3. Applications of Hotspots in PEMS and PMCRs

PEMS is a rapid and nondestructive spectroscopy technique for chemical detection, biosensing, catalysis, and so on. The technique relies on the high density of hotspots to significantly intensify surface-enhanced Raman scattering (SERS) signals, which is expected to achieve single-molecule detection. Based on the SERS spectroscopy technique, nanostructured surfaces and thin films with high density of hotspots as biomarker chips exhibit excellent performance for the specific detection of α-fetoprotein (AFP) and α-fetoprotein-L3 (AFP-L3), which are very promising for the detection of early hepatocellular carcinoma (HCC) markers [38,39].

In addition, based on manipulation of hotspots in nanostructured surfaces and thin films, PMCRs have attracted much attention. Active sites with selectively controlled chemical reactions at the nanometer level can be achieved by manipulating the formation, distribution, and evolution of hotspots in nanostructured surfaces and thin films, which is an easy method to obtain a wide variety of ordered nanostructures and is expected to enhance plasmonic performance.

3. Results and Discussion

3.1. Manipulation of Hotspots in Nanostructured Surfaces and Thin Films

The architectural design and fabrication of multiscale nanostructured surfaces and thin films are carried out by the combination of NSL and magnetron sputtering deposition. An ordered PS colloid sphere array with a size of 500 nm is fabricated by the self-assembly process on Si wafer, which is shown in the schematic diagram in Figure 1a and the scanning electron microscopy (SEM) image in Figure 1c. Hexagonally arranged PS colloid sphere arrays with different sizes and layers can be obtained by a similar process. Using as-prepared PS colloid sphere arrays as a template, when thin films are deposited along the perpendicular direction, two basic nanostructures (nanocap array and triangular-shaped array) are formed, which is the simplest design. By accurately controlling the parameters of the ordered PS colloid sphere array and the processes of magnetron sputtering deposition, basic nanopatterned arrays can be expanded and reinvented to novel nanostructured surfaces. For example, a novel honeycomb nanostructured array can be prepared by selective reactive ion etching (RIE) and glancing angle sputtering with rotation. In the first step, the as-prepared monolayer PS colloid sphere array is etched for different times, which results in six tiny synaptic nanostructures per PS colloid sphere due to the shadow effect, as shown in the schematic diagram in Figure 1b and SEM image in Figure 1d. These tiny synaptic structures among the PS colloid sphere play a key role in the subsequent design of the honeycomb nanostructured array. Then, the Au film is glancing angle sputtered onto the PS colloid sphere array template with tiny synaptic nanostructures by magnetron sputtering (Figure 1e). During the film deposition, the evolution of a well-formed honeycomb nanostructure can be achieved by controlling the rotation speed of the PS colloid sphere array template and film sputtering time, as shown in Figure 1f,g.

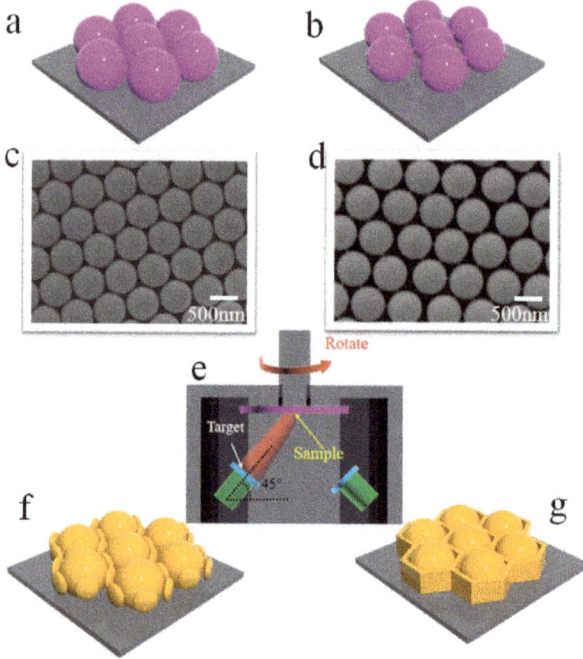

Figure 1. (**a**,**b**) Schematic diagram of hexagonally arranged polystyrene (PS) colloid sphere array and PS colloid sphere array with six tiny synaptic nanostructures after reactive ion etching (RIE) treatment, respectively. (**c**,**d**) SEM image of hexagonally arranged PS colloid sphere array before and after RIE treatment. (**e**) Schematic diagram of the magnetron sputtering chamber. (**f**,**g**) Desired honeycomb nanostructure achieved by controlling rotational speed and film sputtering time. Reproduced with permission from [38] American Chemical Society, 2019.

The morphological features and distribution of hotspots of the honeycomb nanostructures are exactly controlled by RIE time, film sputtering time, and rotation speed during film deposition. To obtain a PS colloid sphere array with tiny synaptic nanostructures and suitable separation, the as-prepared PS colloid sphere array is etched for 120, 180, 240, or 300 s. To optimize the morphological features, we can increase the rotational speed of the etched PS colloid sphere array during the film deposition process (from 10 to 60 rpm), which promotes the rate of film deposition onto the tiny synaptic nanostructure. When the PS colloid sphere array is etched 180 s and the rotation speed is 60 rpm, the honeycomb nanostructure is gradually distinct. The evolution of the honeycomb nanostructured morphology depends on the film deposition time, as shown in Figure 2a–c. When the film deposition time is 5 min, the morphology of the etched PS colloid sphere surface and around the tiny synaptic nanostructures show slight changes (Figure 2a). When the film deposition is increased to 20 min, the synaptic nanostructures around each PS colloid sphere exhibit continuous growth (Figure 2b). When the film deposition time reaches 40 min, a satisfactory honeycomb nanostructure is achieved, where the nanogaps between the sidewalls and the nanocaps are sub-10 nm (Figure 2c). We know that hotspots usually localize in the sharp tips, corners, and sub-10 nm gaps of noble metallic nanostructures, where the coupling EM field is enormously enhanced. FDTD solutions (Lumerical Solutions Inc, Vancouver, BC, Canada) is utilized to simulate the formation, distribution, and evolution of the local EM field in nanostructured surfaces and thin films, where the relevant nanostructural parameters are extracted from the actual prepared patterned nanostructures. Figure 2d–f shows the EM intensity for three fabricated samples (Au film deposition times of 5, 20, and 40 min). The FDTD results indicate that the formation, distribution, and evolution of hotspots can be manipulated by changing the honeycomb nanostructured morphology. The local EM field of the synaptic parts is increased by promoting growth in tiny synaptic structures among the PS colloid sphere (Figure 2d). With an increase in the film deposition time, the hotspots appear and increase in the synaptic nanostructures, as shown in Figure 2e. It is obvious that the local EM field distribution in the nanogaps between the sidewalls and the nanocaps lead to the density of hotspots being predominantly enhanced in a satisfactory honeycomb nanostructure (Figure 2f).

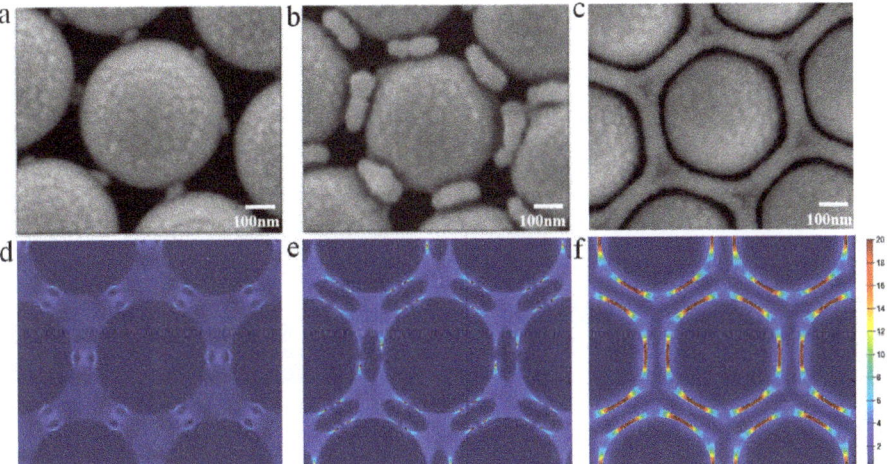

Figure 2. (**a**–**c**) SEM images of the process of honeycomb nanostructure formation with different deposition times (5, 20, and 40 min, respectively). (**d**–**f**) The finite-difference time-domain (FDTD) simulation for sectional views of the local EM field distribution in the process of honeycomb nanostructure formation. Reproduced with permission from [38] American Chemical Society, 2019.

In addition, according to the design requirements, various hybridized, complex, or novel nanostructured surfaces and thin films can be obtained by controlling the preparation strategy. Under a similar process, nanocap star, nanoring nanoparticle, and others nanostructured arrays were achieved in our previous reports [37]. In addition to the size and morphology of nanostructured arrays, the composition of nanostructured surfaces and thin films and the surrounding dielectric environment also play important roles in the manipulation of hotspots. Based on this strategy, noble metals and insulator composites are co-sputtered onto closely ordered PS colloid sphere (200 nm) arrays by magnetron sputtering system (ATC 1800-F, USA AJA). Taking co-deposition of Ag and SiO_2 as an example, the SiO_2-isolated Ag island (SiO_2–Ag) nanocap forms on the PS colloid sphere, as shown in Figure 3a. The transmission electron microscopy (TEM) image shows that the size of the nanogaps between adjacent SiO_2-isolated Ag nanocaps is under sub-10 nm (Figure 3b). The high-resolution transmission electron microscopy (HRTEM) image indicates that the thickness of the amorphous SiO_2 and the size of Ag nanoparticles (AgNPs) are around 2–5 and 5–10 nm, respectively. These amorphous SiO_2 and Ag nanoparticles are intertwined, which helps form nanoscaled surface roughness and more nanogaps (Figure 3c). The corresponding area element analysis mapping of SiO_2–Ag nanocap arrays show that Ag, Si, and O elements are uniformly distributed in the nanocaps, as shown in Figure 3d.

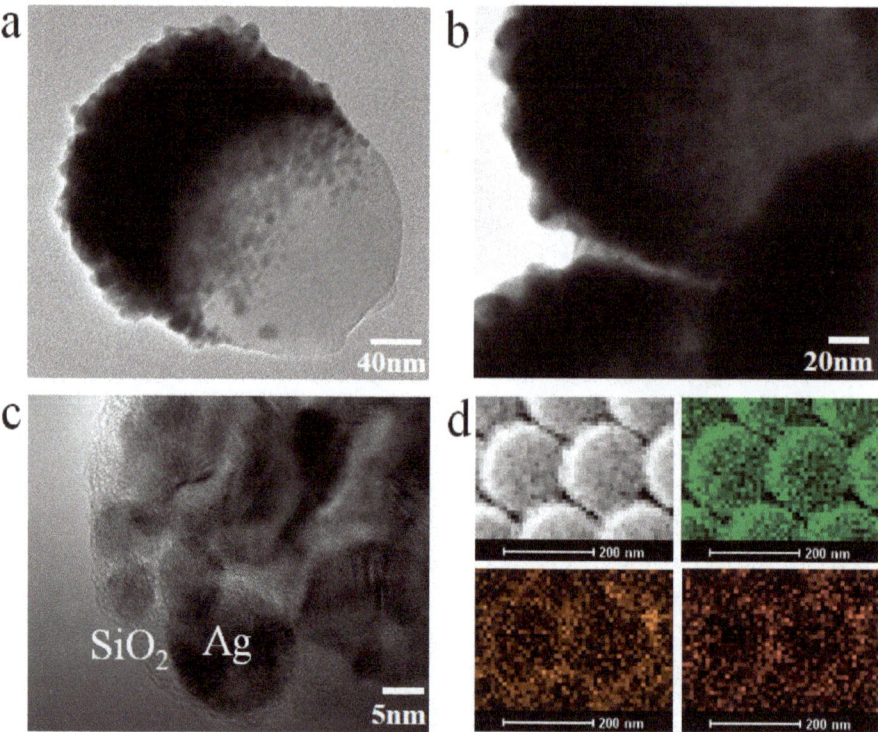

Figure 3. (**a**–**d**) TEM, HRTEM, and element analysis images of the SiO_2–Ag nanocap. Reproduced with permission from [9] American Chemical Society, 2014.

When closely ordered PS colloid sphere arrays are etched, the diameter of PS colloid spheres decreases, and tiny synaptic nanostructures around each PS colloid sphere are observed. After SiO_2–Ag film deposition, the SiO_2–Ag nanocap is still formed on the smaller PS colloid spheres, as shown in Figure 4a. The film preferentially grows around the tiny synaptic nanostructures, which forms a bridge between adjacent nanocaps. With increasing deposition time, the surface roughness of the SiO_2–Ag nanocaps increases, and the nanogaps between the units of bridged knobby units gradually decrease (Figure 4b–d). The results of FDTD simulations indicate that the hotspots where the EM field is coupling are mostly distributed on the surface of the SiO_2-isolated Ag nanoparticles on the nanocaps and the bridges between nanocaps, as shown in Figure 4e. In addition, the trilayer or multilayer

Ag/SiO$_2$ composite shell or 3D pillar-cap arrays also significantly improve the enhancement of the EM field, which manipulates the distribution of hotspots [29,45,51]. The SiO$_2$ addition not only immensely increases the surface roughness of the designed nanostructure surfaces and thin films but also improves the enhancement of the EM field at the nanogaps, which manipulates the formation and evolution of hotspots. The manipulation of hotspots in nanostructured surfaces and thin films has potential applications in the field of plasmonics, including SERS, biomarker chips, mediated chemical reactions, and so on.

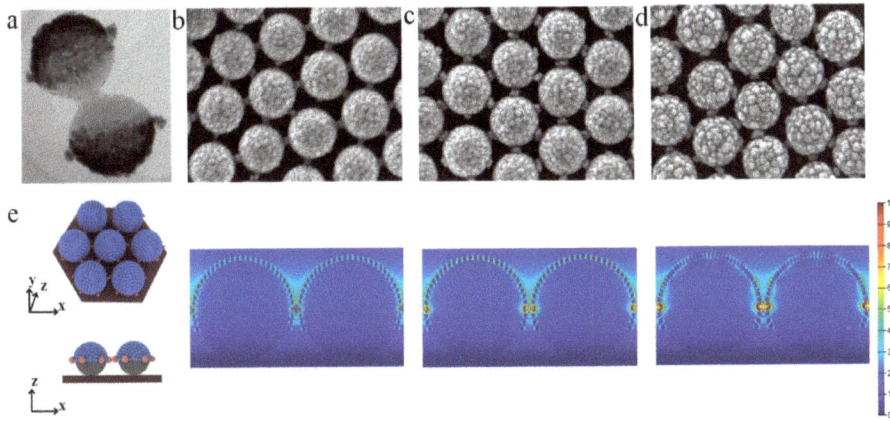

Figure 4. (**a**–**d**) TEM and SEM images of the SiO$_2$–Ag film co-sputtered (5, 15, and 20 min) onto PS colloid sphere array after being etched for 60 s. (**e**) The idealized morphology and simulation results of SiO$_2$–Ag nanocaps in the FDTD. Reproduced from [39] with permission from Elsevier, 2020.

3.2. Applications of Hotspots in Plasmonics

PEMS and PMCRs are two important branches in plasmonics. PEMS includes fluorescence spectrum, infrared spectrum, and SERS. Among the three spectra, SERS is the most promising spectroscopy technique, which can achieve obviously enhanced Raman signals by hotspots in nanostructured surfaces and thin films. As mentioned above, Au nanohoneycomb and SiO$_2$–Ag nanocap arrays show typical SERS spectra when 4-mercaptobenzoic acid (4-MBA) is used as a probe. Figure 5a–b show that SERS peaks at about 1575, 1073, and 1173 cm^{-1} are assigned to the aromatic ring vibrations and the C–H deformation vibration modes, respectively [37]. The SERS intensity of 4-MBA increases with the manipulation of hotspots in nanostructured surfaces and thin films and obtains the highest enhancement, which is in good agreement with the FDTD simulation.

Based on the SERS technique, the biological and biomedical detection of some mortal diseases has been achieved by determining changes in the shift and intensity of the SERS signals. For instance, using SiO$_2$–Ag nanocap arrays with high density of hotspots as biomarker chips, the early diagnosis for HCC can be detected based on the analysis of the shift in characteristic peaks of the probe molecule 4-MBA and AFP-L3. The whole preparation and immune process of biomarker chips for the detection of HCC is shown in Figure 6a. First, the biomarker chip, which is composed of SiO$_2$–Ag nanocap arrays with bridges, is immersed in the 4-MBA, 1-(3-(dimethylamino)propyl)-3-ethylcarbodimide hydrochloride (EDC), and N-hydroxysuccinimide (NHS) solutions. Then, the 4-MBA-derived coupling agent is generated using EDC–NHS, which is used to bind the anti-AFP antibody. After the reaction, bovine serum albumin (BSA) is added into the mixed solution to block unconnected anti-AFP antibodies.

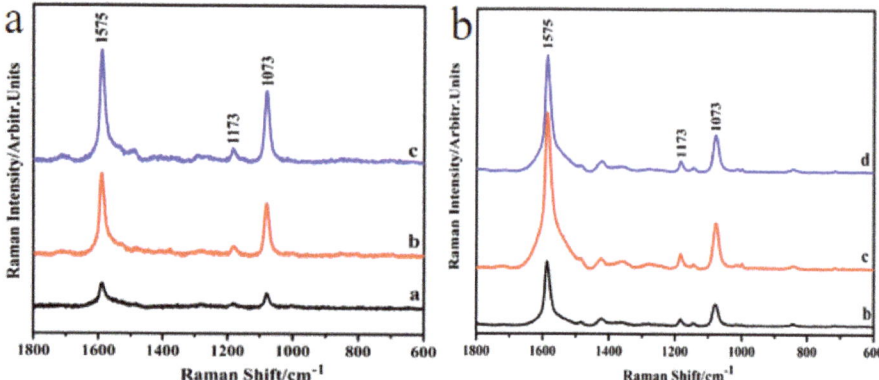

Figure 5. (a,b) The surface-enhanced Raman scattering (SERS) spectrum of 4-mercaptobenzoic acid (4-MBA) for the Au nanohoneycomb and SiO$_2$–Ag nanocap arrays, which correspond to Figures 2a–c and 4b–d, respectively. Reproduced with permission from [38] American Chemical Society, 2019; Reproduced with permission from [9] American Chemical Society, 2014.

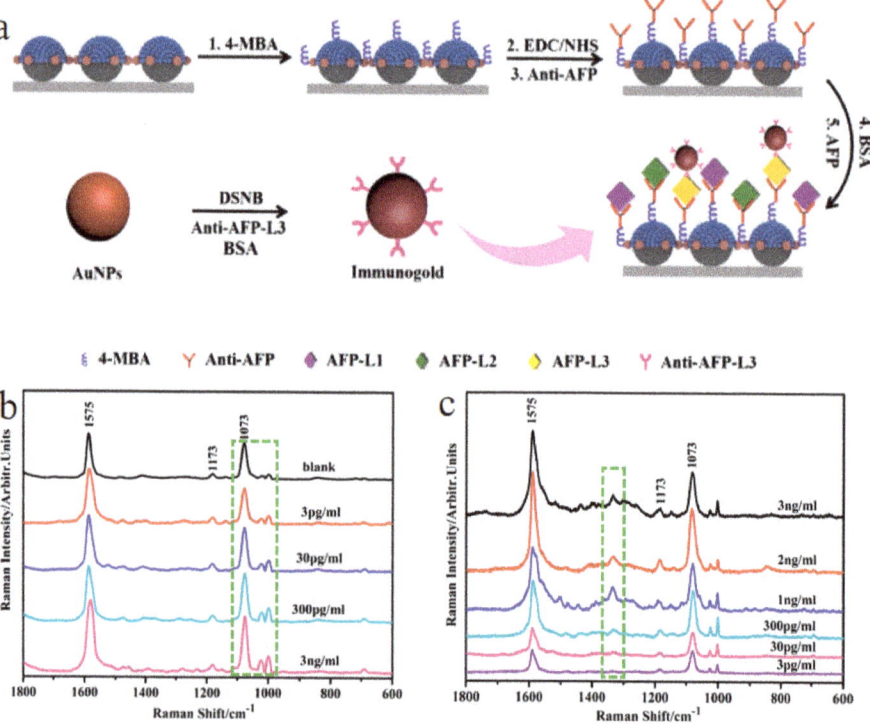

Figure 6. (a) The preparation schematic diagram of SiO$_2$–Ag nanocap arrays as a biomarker chip. (b,c) The SERS spectrum of the biomarker chip for different α-fetoprotein (AFP) and α-fetoprotein-L3 (AFP-L3) concentrations, respectively. Reproduced from [39] with permission from Elsevier, 2020.

In addition, the anti-AFP is diluted and used as a blank contrast sample. Antigens with different concentrations (3, 30, 300, and 3000 pg/mL) are added and allowed to react with the biomarker chip in the centrifuge tubes. After the process, antibody-capturing chips are made and characterized by SERS. The SERS spectrum of 4-MBA connected with different AFP concentrations are shown in Figure 6b. When the concentration of the AFP increases, the peaks of 4-MBA around 1073 cm^{-1} shifts to the left and the intensity of the peaks at 998 cm^{-1} are enhanced, as shown by the dashed frame in Figure 6b. The changes in the two peaks confirm the success of the MBA-based antibody absorption. Subsequently, the preparation of immunogold and immunological recognition are implemented to detect HCC by analyzing the ratio of AFP-L3 to total AFP. For analyzing AFP-L3, 5,5′-dithiobis (succinimidyl-2-nitrobenzoate) (DSNB) is used as a probe between the antibody and the AuNPs. Colloidal gold is fabricated by the Lee and Meisel approach, which is added to the DSNB acetonitrile solution. Next, the anti-AFP-L3 is put in and stored at room temperature. After the reaction, BSA and borate buffer are added into the solution for later use. Finally, the biomarker chip is formed after being immersed in solutions of different concentrations (3, 2, 1 ng/mL and 300, 30, 3 pg/mL) to ensure sufficient immunological recognition. Figure 6c shows the SERS spectrum of the immunogold-decorated biomarker chip with different AFP-L3 concentrations. The SERS peak at 1331 cm^{-1} is used as a characteristic immunogold signal in the dashed frame in Figure 6c, which reflects the degree of coupling between AuNPs. With the methods mentioned above, the detection limits for AFP and AFP-L3 are below 3 pg/mL, which proves that the designed nanostructured surfaces and thin films with ordered hotspots is of great significance for the early detection of HCC, clinical application, and SERS immune detection [38,39].

PMCRs can be designed based on manipulation of hotspots in nanostructured surfaces and thin films. As we know, the hotspots of Au nanobowl arrays are located on the edge of the Au nanobowl, as shown in Figure 7a. Due to the formation, distribution, and evolution of hotspots under the photoinduced effect of the enhanced EM field, the Au nanobowl arrays induce a photoreaction, leading to accelerated chemical reaction on defined positions. Interestingly, the size and position of the AgNPs are precisely controlled by polarized light and reaction time, as shown in Figure 7b–d. When using vertically circular polarized light incident to the surface of the Au nanobowl arrays, a six-axis symmetric patterned arrays of AgNPs growth is achieved. Furthermore, three-axis symmetric nanostructured arrays are obtained using linearly polarized oblique waves with a incidence angle of 50 degrees. The manipulation of hotspots can be used to accurately control a chemical reaction at the nanometer level, which has significant applications [41–48,53,55].

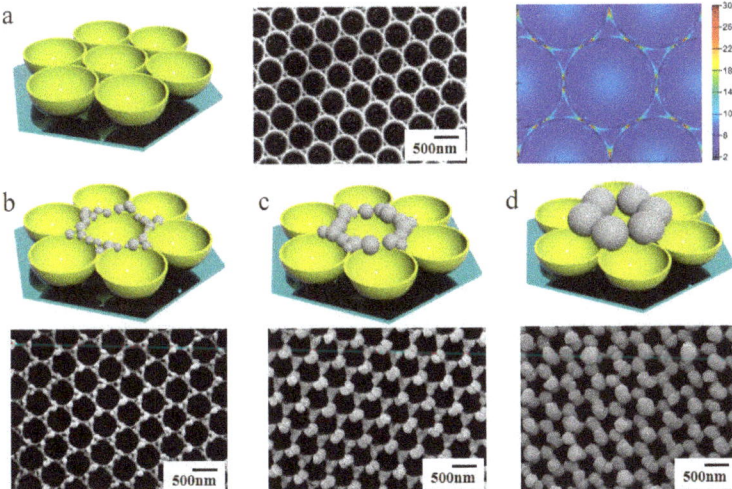

Figure 7. (a) SEM image and the distribution of hotspots for the Au nanobowl arrays. (b–d) SEM images of the Au nanobowl arrays with Ag nanoparticles (AgNPs) accurately grown at the defined hotspot location. Reproduced from [13] with permission from AIP Publishing, 2019.

NSL-based nanostructured surfaces and thin films have also been focused on various fields, including superhydrophobicity [62], protein patterning [63], magnetization reversal [64], solar cells [65], light trapping enhancement [66], resonant optical transmission [31], flexible broadband antireflective coatings [67], flexibly tunable smart displays, and many others [68], which show more novel and interesting properties.

4. Conclusions and Outlook

In summary, various hybrid and even complex nanostructured surfaces and thin films can be designed and achieved by the combination of NSL, physical vapor deposition technique, etching, transfer, chemical reactions, or their combination steps. The formation, distribution, and evolution of hotspots can be manipulated by fabricating novel nanohoneycomb and SiO_2–Ag nanocap arrays to control the enhancement of the EM field, which has potential applications in PEMS and PMCRs. In particular, detection of HCC based on SERS and controlling the growth locations of AgNPs are attracting more and more attention, which is expected to enhance the sensing performance. However, solving the defect formation during the self-assembly process for nanostructure-based devices is necessary in NSL technology. The notable result on defect-free PS colloid sphere arrays over a large area (36 wafers and 1 m^2) has been demonstrated using the micropropulsive injection method to achieve high-throughput (6 × 6 wafers) periodic surface nanotexturing [58]. Recently, NSL-based nanostructured surfaces and thin films with functional materials have shown significant application in emerging magnetic skyrmion-based spintronic devices [69]. Thus, further functionalization of nanostructured surfaces and thin films with more ingenious designs are expected to allow for unprecedented versatility of NSL in a broad range of applications.

Author Contributions: Y.Z. and J.W. conceived and designed the experiments; X.Z. (Xiaoyu Zhao) and Y.W. contributed significantly to the analysis and manuscript preparation; X.Z. (Xiaolong Zhang), A.Z., M.C. and Q.Z. performed the experiments and analyzed the data. All authors have read and agreed to the published version of the manuscript.

Funding: This research received no external funding.

Acknowledgments: This work is supported by the Fundamental Research Funds for the Provincial Universities of Zhejiang (GK20997299001) and the National Natural Science Foundation of China (No. 51901060, 61575080, and 61675090).

Conflicts of Interest: The authors declare no conflict of interest.

References

1. Kuznetsov, A.I.; Miroshnichenko, A.E.; Brongersma, M.L.; Kivshar, Y.S.; Luk'yanchuk, B. Optically resonant dielectric nanostructures. *Science* **2016**, *354*, 2472. [CrossRef]
2. Ai, B.; Yu, Y.; Möhwald, H.; Zhang, G.; Yang, B. Plasmonic films based on colloidal lithography. *Adv. Colloid Interface Sci.* **2014**, *206*, 5. [CrossRef]
3. Fan, F.R.; Tang, W.; Wang, Z.L. Flexible nanogenerators for energy harvesting and self-powered electronics. *Adv. Mater.* **2016**, *28*, 4283–4305. [CrossRef]
4. Wang, Z.Y.; Ai, B.; Möhwald, H.; Zhang, G. Colloidal lithography meets plasmonic nanochemistry. *Adv. Opt. Mater.* **2018**, *6*, 1800402. [CrossRef]
5. Zhu, T.; Wang, H.; Zang, L.B.; Jin, S.L.; Guo, S.; Park, E.; Mao, Z.; Jung, Y.M. Flexible and reusable Ag coated TiO_2 nanotube arrays for highly sensitive SERS detection of formaldehyde. *Molecules* **2020**, *25*, 1199. [CrossRef]
6. Pellacani, P.; Morasso, C.; Picciolini, S.; Gallach, D.; Fornasari, L.; Marabelli, F.; Silvan, M.M. Plasma fabrication and SERS functionality of gold crowned silicon submicrometer pillars. *Materials* **2020**, *13*, 1244. [CrossRef]
7. Ma, L.W.; Wang, J.K.; Huang, H.C.; Zhang, Z.J.; Li, X.G.; Fan, Y. Simultaneous thermal stability and ultrahigh sensitivity of heterojunction SERS substrates. *Nanomaterials* **2019**, *9*, 830. [CrossRef]
8. Willets, K.A.; van Duyne, R.P. Localized surface plasmon resonance spectroscopy and sensing. *Ann. Rev. Phys. Chem.* **2007**, *58*, 267. [CrossRef]

9. Wang, Y.X.; Zhao, X.Y.; Chen, L.; Chen, S.; Wei, M.B.; Gao, M.; Zhao, Y.; Wang, C.; Qu, X.; Zhang, Y.J.; et al. Ordered nanocap array composed of SiO$_2$-isolated Ag islands as SERS platform. *Langmuir* **2014**, *30*, 15285. [CrossRef]
10. Chen, Y.; Shu, Z.; Feng, Z.; Kong, L.; Liu, Y.; Duan, H. Reliable patterning, transfer printing and post-assembly of multiscale adhesion-free metallic structures for nanogap device applications. *Adv. Funct. Mater.* **2020**, 2002549. [CrossRef]
11. Chen, Y.; Bi, K.; Wang, Q.; Zheng, M.; Liu, Q.; Han, Y.; Yang, J.; Chang, S.; Zhang, G.; Duan, H. Rapid focused ion beam milling based fabrication of plasmonic nanoparticles and assemblies via "Sketch and Peel" strategy. *ACS Nano* **2016**, *10*, 11228–11236. [CrossRef]
12. Chen, Y.; Xiang, Q.; Li, Z.; Wang, Y.; Meng, Y.; Duan, H. "Sketch and Peel" lithography for high-resolution multiscale patterning. *Nano Lett.* **2016**, *16*, 3253–3259. [CrossRef]
13. Zhao, X.Y.; Wen, J.H.; Li, L.W.; Wang, Y.X.; Wang, D.H.; Chen, L.; Zhang, Y.J.; Du, Y.W. Architecture design and applications of nanopatterned arrays based on colloidal Lithography. *J. Appl. Phys.* **2019**, *126*, 141101. [CrossRef]
14. Stiles, P.L.; Dieringer, J.A.; Shah, N.C.; van Duyne, R.P. Surface-enhanced Raman spectroscopy. *Ann. Rev. Anal. Chem.* **2011**, *1*, 601. [CrossRef]
15. Farcău, C.; Aştilean, S. Silver half-shell arrays with controlled plasmonic response for fluorescence enhancement optimization. *Appl. Phys. Lett.* **2009**, *95*, 193110.
16. Kreno, L.E.; Greeneltch, N.G.; Farha, O.K.; Hupp, J.T.; van Duyne, R.P. SERS of molecules that do not adsorb on Ag surfaces: A metal–organic framework-based functionalization strategy. *Analyst* **2014**, *139*, 4073. [CrossRef]
17. Wang, Y.X.; Zhao, X.Y.; Gao, W.T.; Chen, L.; Chen, S.; Wei, M.B.; Gao, M.; Wang, C.; Zhang, Y.J.; Yang, J.H. Au/Ag bimetal nanogap arrays with tunable morphologies for surface-enhanced Raman scattering. *RSC Adv.* **2015**, *5*, 7454. [CrossRef]
18. Zhang, Y.J.; Wang, C.; Wang, J.P.; Chen, L.; Li, J.; Liu, Y.; Zhao, X.Y.; Wang, Y.X.; Yang, J.H. Nanocap array of Au:Ag composite for surface-enhanced Raman scattering. *Spectrochim. Acta Part Mol. Biomol. Spectrosc.* **2016**, *152*, 461. [CrossRef]
19. Tabatabaei, M.; Sangar, A.; Kazemi-Zanjani, N.; Torchio, P.; Merlen, A.; Lagugne-Labarthet, F. Optical properties of silver and gold tetrahedral nanopyramid arrays prepared by nanosphere lithography. *J. Phys. Chem. C* **2013**, *117*, 14778. [CrossRef]
20. Zrimsek, A.B.; Henry, A.-I.; van Duyne, R.P. Single molecule Surface-enhanced Raman spectroscopy without nanogaps. *J. Phys. Chem. Lett.* **2013**, *4*, 3206. [CrossRef]
21. Jensen, T.R.; Malinsky, M.D.; Haynes, C.L.; van Duyne, R.P. Nanosphere Lithography: Tunable localized surface plasmon resonance spectra of silver nanoparticles. *J. Phys. Chem. B* **2000**, *104*, 10549. [CrossRef]
22. Dickreuter, S.; Gleixner, J.; Kolloch, A.; Boneberg, J.; Scheer, E.; Leiderer, P. Mapping of plasmonic resonances in nanotriangles. *Beil. J. Nanotechnol.* **2013**, *4*, 588. [CrossRef]
23. Vogel, N.; Belisle, R.A.; Hatton, B.; Wong, T.S.; Aizenberg, J. Transparency and damage tolerance of patternable omniphobic lubricated surfaces based on inverse colloidal monolayers. *Nat. Commun.* **2013**, *4*, 2176. [CrossRef]
24. Yang, S.K.; Lapsley, M.L.; Cao, B.Q.; Zhao, C.L.; Zhao, Y.H.; Hao, Q.Z.; Kiraly, B.; Scott, J.; Li, W.Z.; Wang, L.; et al. Large-scale fabrication of three-dimensional surface patterns using template-defined electrochemical deposition. *Adv. Funct. Mater.* **2013**, *23*, 720. [CrossRef]
25. Yang, S.K.; Cao, B.Q.; Kong, L.; Wang, Z.Y. Template-directed dewetting of a gold membrane to fabricate highly SERS-active substrates. *J. Mater. Chem.* **2011**, *21*, 14031. [CrossRef]
26. Ai, B.; Basnet, P.; Larson, S.; Ingram, W.; Zhao, Y.P. Plasmonic sensor with high figure of merit based on differential polarization spectra of elliptical nanohole array. *Nanoscale* **2010**, *9*, 14710. [CrossRef]
27. Lee, S.H.; Bantz, K.C.; Lindquist, N.C.; Oh, S.H.; Haynes, C.L. Self-assembled plasmonic nanohole arrays. *Langmuir* **2009**, *25*, 13685. [CrossRef]
28. Pisco, M.; Galeotti, F.; Quero, G.; Grisci, G.; Micco, A.; Mercaldo, L.V.; Veneri, P.D.; Cutolo, A.; Cusano, A. Nanosphere lithography for optical fiber tip nanoprobes. *Light Sci. Appl.* **2017**, *6*, 1. [CrossRef]
29. Wang, Y.X.; Zhang, M.N.; Yan, C.; Chen, L.; Liu, Y.; Li, J.; Zhang, Y.J.; Yang, J.H. Pillar-cap shaped arrays of Ag/SiO$_2$ multilayers after annealing treatment as a SERS-active substrate. *Colloids Surf. A Phys. Eng. Asp.* **2016**, *506*, 96. [CrossRef]

30. Gao, R.X.; Zhang, Y.J.; Zhang, F.; Guo, S.; Wang, Y.X.; Chen, L.; Yang, J.H. SERS polarization-dependent effects for an ordered 3D plasmonic tilted silver nanorod array. *Nanoscale* **2018**, *10*, 8106. [CrossRef]
31. Ai, B.; Yu, Y.; Möhwald, H.; Wang, L.M.; Zhang, G. Resonant optical transmission through topologically continuous films. *ACS Nano* **2014**, *8*, 1566. [CrossRef]
32. Ai, B.; Gu, P.P.; Wang, Z.Y.; Möhwald, H.; Wang, L.M.; Zhang, G. Light trapping in plasmonic nanovessels. *Adv. Opt. Mater.* **2017**, *5*, 1600980. [CrossRef]
33. Ai, B.; Möhwald, H.; Zhang, G. Smart pattern display by tuning the surface plasmon resonance of hollow nanocone arrays. *Nanoscale* **2015**, *7*, 11525. [CrossRef]
34. Yang, S.K.; Sun, N.; Stogin, B.B.; Wang, J.; Huang, Y.; Wong, T.S. Ultra-antireflective synthetic brochosomes. *Nat. Commun.* **2017**, *8*, 1285. [CrossRef]
35. Yu, Y.; Zhou, Z.W.; Möhwald, H.; Ai, B.; Zhao, Z.Y.; Ye, S.S.; Zhang, G. Distorted colloidal arrays as designed template. *Nanotechnology* **2015**, *26*, 035301. [CrossRef]
36. Ho, C.C.; Zhao, K.; Lee, T.Y. Quasi-3D gold nanoring cavity arrays with high-density hot-spots for SERS applications via nanosphere lithography. *Nanoscale* **2014**, *6*, 8606. [CrossRef]
37. Zhao, X.Y.; Wen, J.H.; Zhang, M.N.; Wang, D.H.; Wang, Y.X.; Chen, L.; Zhang, Y.J.; Yang, J.H.; Du, Y.W. Design of hybrid nanostructural arrays to manipulate SERS-active substrates by nanosphere lithography. *ACS Appl. Mater. Interfaces* **2017**, *9*, 7710. [CrossRef]
38. Zhu, A.N.; Zhao, X.Y.; Cheng, M.Y.; Chen, L.; Wang, Y.X.; Zhang, X.L.; Zhang, Y.J.; Zhang, X.F. Nanohoneycomb Surface-enhanced Raman spectroscopy-active chip for the determination of biomarkers of hepatocellular carcinoma. *ACS Appl. Mater. Interfaces* **2019**, *11*, 44617–44623. [CrossRef]
39. Cheng, M.Y.; Zhang, F.; Zhu, A.N.; Zhang, X.L.; Wang, Y.X.; Zhao, X.Y.; Chen, L.; Hua, Z.; Zhang, Y.J.; Zhang, X.F. Bridging the neighbor plasma coupling on curved surface array for early hepatocellular carcinoma detection. *Sens. Actuators B Chem.* **2020**, *309*, 127759. [CrossRef]
40. Zhu, K.; Wang, Z.Y.; Zong, S.F.; Liu, Y.; Yang, K.; Li, N.; Wang, Z.L.; Li, L.; Tang, H.L.; Cui, Y.P. Hydrophobic plasmonic nanoacorn array for a label-free and uniform SERS-based biomolecular assay. *ACS Appl. Mater. Interfaces* **2020**, *12*, 29917–29927.
41. Yang, S.K.; Lei, Y. Recent progress on surface pattern fabrications based on monolayer colloidal crystal templates and related applications. *Nanoscale* **2011**, *3*, 2768. [CrossRef]
42. Ai, B.; Möhwald, H.; Wang, D.Y.; Zhang, G. Advanced colloidal lithography beyond surface patterning. *Adv. Mater. Interfaces* **2016**, *4*, 1600271. [CrossRef]
43. Chen, L.; Sun, H.H.; Zhao, Y.; Gao, R.X.; Wang, Y.X.; Liu, Y.; Zhang, Y.J.; Hua, Z.; Yang, J.H. Iron layer-dependent surface-enhanced Raman scattering of hierarchical nanocap arrays. *Appl. Surf. Sci.* **2017**, *423*, 1124. [CrossRef]
44. Zhang, X.Y.; Chen, L.; Wang, Y.X.; Zhang, Y.J.; Yang, J.H.; Choi, H.C.; Jung, Y.M. Design of tunable ultraviolet (UV) absorbance by controlling the Ag-Al co-sputtering deposition. *Spectrochim. Acta Part A Mol. Biomol. Spectrosc.* **2018**, *197*, 37. [CrossRef]
45. Zhang, Y.J.; Sun, H.H.; Gao, R.X.; Zhang, F.; Zhu, A.N.; Chen, L.; Wang, Y.X. Facile SERS-active chip (PS@Ag/SiO$_2$/Ag) for the determination of HCC biomarker. *Sens. Actuators B Chem.* **2018**, *272*, 34. [CrossRef]
46. Yang, S.K.; Hricko, P.J.; Huang, P.H.; Li, S.X.; Zhao, Y.H.; Xie, Y.L.; Guo, F.; Wang, L.; Huang, T.J. Superhydrophobic surface enhanced Raman scattering sensing using janus particle arrays realized by site-specific electrochemical growth. *J. Mater. Chem. C* **2014**, *2*, 542. [CrossRef]
47. Correia-Ledo, D.; Gibson, K.F.; Dhawan, A.; Couture, M.; Vo-Dinh, T.; Graham, D.; Masson, J.F. Assessing the location of surface plasmons over nanotriangle and nanohole arrays of different size and periodicity. *J. Phys. Chem. C* **2012**, *116*, 6884. [CrossRef]
48. Ai, B.; Wang, L.M.; Möhwald, H.; Yu, Y.; Zhang, G. Asymmetric half-cone/nanohole array films with structural and directional reshaping of extraordinary optical transmission. *Nanoscale* **2014**, *6*, 8997. [CrossRef]
49. Haynes, C.L.; van Duyne, R.P. Nanosphere lithography: A versatile nanofabrication tool for studies of size-dependent nanoparticle optics. *J. Phys. Chem. B* **2001**, *105*, 5599. [CrossRef]
50. Wang, Y.X.; Yan, C.; Chen, L.; Zhang, Y.J.; Yang, J.H. Controllable charge transfer in Ag-TiO$_2$ composite structure for SERS application. *Nanomaterials* **2017**, *7*, 159. [CrossRef]
51. Zhang, F.; Wang, Y.X.; Zhang, Y.J.; Chen, L.; Liu, Y.; Yang, J.H. Ag nanotwin-assisted grain growth-induced by stress in SiO$_2$/Ag/SiO$_2$ nanocap arrays. *Nanomaterials* **2018**, *8*, 436. [CrossRef]

52. Zhang, F.; Guo, S.; Liu, Y.; Chen, L.; Wang, Y.X.; Gao, R.X.; Zhu, A.N.; Zhang, X.L.; Zhang, Y.J. Controlling the 3D electromagnetic coupling in co-sputtered Ag-SiO$_2$ nanomace arrays by lateral sizes. *Nanomaterials* **2018**, *8*, 493. [CrossRef]
53. Zhu, A.N.; Gao, R.X.; Zhao, X.Y.; Zhang, F.; Zhang, X.Y.; Yang, J.H.; Zhang, Y.J.; Chen, L.; Wang, Y.X. Site-selective growth of Ag nanoparticles controlled by localized surface plasmon resonance of nanobowl arrays. *Nanoscale* **2019**, *11*, 6576. [CrossRef]
54. Zhang, F.; Zhang, Y.J.; Wang, Y.X.; Zhao, X.Y. Composition-dependent LSPR shifts for co-sputtered TiS$_2$-Ag. *Opt. Commun.* **2020**, *473*, 125935. [CrossRef]
55. Zhu, Q.; Zhang, X.L.; Wang, Y.X.; Zhu, A.N.; Gao, R.X.; Zhao, X.Y.; Zhang, Y.J.; Chen, L. Controlling the growth locations of Ag nanoparticles at nanoscale by shifting LSPR hotspots. *Nanomaterials* **2019**, *9*, 1553. [CrossRef]
56. Ellinas, K.; Smyrnakis, A.; Malainou, A.; Tserepi, A.; Gogolides, E. "Mesh-assisted" colloidal lithography and plasma etching: A route to large-area, uniform, ordered nano-pillar and nanopost fabrication on versatile substrates. *Microelectron. Eng.* **2011**, *88*, 2547. [CrossRef]
57. Yang, S.M.; Jang, S.G.; Choi, D.G.; Kim, S.; Yu, H.K. Nanomachining by Colloidal Lithography. *Small* **2006**, *2*, 458. [CrossRef]
58. Gao, P.Q.; He, J.; Zhou, S.Q.; Yang, X.; Li, S.Z.; Sheng, J.; Wang, D.; Yu, T.B.; Ye, J.C. Large-area nanosphere self-assembly by a micro-propulsive injection method for high throughput periodic surface nanotexturing. *Nano Lett.* **2015**, *15*, 4591. [CrossRef]
59. Li, Y.; Cai, W.P.; Duan, G.T. Ordered micro/nanostructured arrays based on the monolayer colloidal crystals. *Chem. Mater.* **2008**, *20*, 615. [CrossRef]
60. Skehan, C.; Ai, B.; Larson, S.R.; Stone, K.M.; Dennis, W.M.; Zhao, Y.P. Plasmonic and SERS performances of compound nanohole arrays fabricated by shadow sphere lithography. *Nanotechnology* **2018**, *29*, 095301. [CrossRef]
61. Zhang, G.; Wang, D.Y.; Möhwald, H. Patterning microsphere surfaces by templating colloidal crystals. *Nano Lett.* **2005**, *5*, 143. [CrossRef]
62. Ellinas, K.; Tserepi, A.; Gogolides, E. From superamphiphobic to amphiphilic polymeric surfaces with ordered hierarchical roughness fabricated with colloidal lithography and plasma nanotexturing. *Langmuir* **2011**, *27*, 3960. [CrossRef]
63. Malainou, A.; Tsougeni, K.; Ellinas, K.; Petrou, P.S.; Constantoudis, V.; Sarantopoulou, E.; Awsiuk, K.; Bernasik, A.; Budkowski, A.; Markou, A.; et al. Plasma-assisted nanoscale protein patterning on si substrates via colloidal lithography. *J. Phys. Chem. A* **2013**, *117*, 13743. [CrossRef]
64. Markou, A.; Beltsios, K.G.; Gergidis, L.N.; Panagiotopoulos, I.; Bakas, T.; Ellinas, K.; Tserepi, A.; Stoleriu, L.; Tanasa, R.; Stancu, A. Magnetization reversal in triangular L$_{10}$-FePt nanoislands. *J. Magn. Magn. Mater.* **2013**, *344*, 224. [CrossRef]
65. Zhou, S.Q.; Yang, Z.H.; Gao, P.Q.; Li, X.F.; Yang, X.; Wang, D.; He, J.; Ying, Z.Q.; Ye, J.C. Wafer-scale integration of inverted nanopyramid arrays for advanced light trapping in crystalline silicon thin film solar cells. *Nanoscale Res. Lett.* **2016**, *11*, 194. [CrossRef]
66. Yang, X.; Zhou, S.Q.; Wang, D.; He, J.; Zhou, J.; Li, X.F.; Gao, P.Q.; Ye, J.C. Light trapping enhancement in a thin film with 2D conformal periodic hexagonal arrays. *Nanoscale Res. Lett.* **2015**, *10*, 284. [CrossRef]
67. Toma, M.; Loget, G.; Corn, R.M. Fabrication of broadband antireflective plasmonic gold nanocone arrays on flexible polymer films. *Nano Lett.* **2013**, *13*, 6164. [CrossRef]
68. Fan, W.; Zeng, J.; Gan, Q.Q.; Ji, D.X.; Song, H.M.; Liu, W.Z.; Shi, L.; Wu, L.M. Iridescence-controlled and flexibly tunable retroreflective structural color film for smart displays. *Sci. Adv.* **2019**, *5*, 8755. [CrossRef]
69. Wang, Y.; Wang, L.; Xia, J.; Lai, Z.; Tian, G.; Zhang, X.; Hou, Z.P.; Gao, X.S.; Mi, W.B.; Feng, C.; et al. Electric-field-driven non-volatile multi-state switching of individual skyrmions in a multiferroic heterostructure. *Nat. Commun.* **2020**, *11*, 3577. [CrossRef]

© 2020 by the authors. Licensee MDPI, Basel, Switzerland. This article is an open access article distributed under the terms and conditions of the Creative Commons Attribution (CC BY) license (http://creativecommons.org/licenses/by/4.0/).

MDPI
St. Alban-Anlage 66
4052 Basel
Switzerland
Tel. +41 61 683 77 34
Fax +41 61 302 89 18
www.mdpi.com

Nanomaterials Editorial Office
E-mail: nanomaterials@mdpi.com
www.mdpi.com/journal/nanomaterials

www.ingramcontent.com/pod-product-compliance
Lightning Source LLC
LaVergne TN
LVHW070646100526
838202LV00013B/893